U0023327
Cook 50

Cook 50

Cook 50

Cook 50

冰涼果凍、冰寶、涼糕、冰淇淋、慕斯＋沁涼小菜

JELLO · DESSERT · CHILLED SIDE DISH

心凍小品百分百

梁淑嫈 著
by Emily Liang

朱雀文化事業有限公司

Preface

讓人胃口大開的心凍小品

想寫一本這樣的食譜已經好多年了，早期吉利丁這個名詞對於許多人來說還是非常陌生的，市面上很難得看得到，常常會接到讀者的電話詢問；漸漸的，這幾年西風東漸，突然之間書店裡出現了好多西點食譜，西點材料專賣店也一家一家開，吉利丁在西點裡廣泛的被運用，因此現在已經很容易購買得到了，而一般人也都知道吉利丁的用途。

本來我只是計畫寫一本如何使用吉利丁的食譜，但在製作這本書的過程中，不斷地回想起自己小時候酷愛吃的，媽媽親手做的牛肉凍，還有夏天和姊妹們邊做邊玩邊吃的雞蛋布丁，就決定把目前坊間常用到的各種天然凝固劑放進書裡，設計出各式各樣的甜、鹹小品，所以你會發現從最傳統的洋菜粉、布丁粉，到葛粉、地瓜粉，甚至豬皮都可以烹調出各種食物，不僅可製作甜點、果凍，還可以做出各式各樣冰涼的菜餚。

在炎熱的夏日或家人沒有食欲時，不妨在烹飪上來點變化，一客冰涼的甜品降低了不少暑氣，而一道沁涼的菜餚則讓人胃口大開；對家有小朋友的主婦來說，我相信只要以果凍方式呈現的點心或菜餚，都會讓孩子們食欲大增。生活是需要多點創意的，希望這本食譜的誕生，讓你的生活增添更多情趣。

I have been thinking about writting a book like this for many years. Just a few years ago, even the word "gelatin" was a mystery to most people. It was difficult to find it in the market. Frequently readers would call and ask just what it was. Over the last few years, as western influences have gradually made themselves felt here in the east, suddenly bookstore shelves are packed with western cookbooks. Western ingredient shops are springing up everywhere. In western desserts, gelatin is a common sight. As a result, it is easy to purchase gelatin, and ordinary people are familiar with its use.

Originally I thought I'd just write a book about how to use gelatin. However, as I was writing this book, I kept thinking back to the things I liked to eat when I was little, my mother's beef jello or my sisters and I ate egg pudding as we played. Finally, I decided to put in the book all the natural thickening agents available in the market. I designed this book to contain a tremendous variety of dishes, from sweet desserts to meat jello fit for a feast. You will discover that traditional thickening agents such as agar-agar, pudding powder, arrowroot powder, yam flour, and even pork skin can be used to cook practically any dish. Not only can desserts be made from them, but an endless number of cold dishes as well.

On hot summer days or when the family just doesn't feel like eating, why not give the family a change of pace? A chilled, sweet dish can stimulate the appetite. As for housewife with children, I believe that gelatin-based sweets and dishes would increase the appetite of kids. Life requires a little creativity now and then. I hope that this book will be an enriching and energizing experience for you and your cooking.

梁淑嫈
Emily Liang

目錄 CONTENTS

2 讓人胃口大開的心凍小品
PREFACE

6 如何做出心凍美食？
HOW TO USE THICKENERS TO MAKE CHILLED DISHES?

冰涼甜點

12 洛神仙楂凍
ROSELLA WITH CRATAEGUS JELLO

14 什錦水果凍
MIXED FRUIT JELLO

16 五彩優格果凍
YOGURT FRUIT JELLO

18 番茄果凍
TOMATO JELLO

20 綠茶亞答子凍
GREEN TEA WITH YATATZ JELLO

22 蜜汁胡蘿蔔凍
CARROT WITH HONEY JELLO

24 南瓜布丁
PUMPKIN PUDDING

26 雞蛋布丁
EGG PUDDING

28 草莓涼糕
COLD STRAWBERRY CAKE

30 綠豆涼糕
COLD MUNG BEAN CAKE

32 紅豆軟糕
SOFT RED BEAN CAKE

34 西米布丁
SAGO PUDDING

36 椰汁雪花糕
COCONUT CAKE

38 桂花蓮子凍
OSMANTHUS WITH LOTUS SEED JELLO

40 杏仁豆腐
ALMOND TOFU CAKE

42 草莓冰寶
STRAWBERRY SHERBET

43 芒果冰寶
MANGO SHERBET

CONTENTS

44 可樂冰寶
COKE SHERBET
46 香草冰淇淋
VANILLA ICE CREAM
48 核桃巧克力冰淇淋
WALNUT CHOCOLATE
ICE CREAM
50 香橙軟糖
SOFT ORANGE CANDIES
52 葡萄軟糖
SOFT GRAPE CANDIES
54 枸杞馬蹄糕
LYCIUM BERRY AND
WATER CHESTNUT CAKE
56 芋頭涼糕
COLD TARO CAKE
58 綠豆蓮藕糕
MUNG BEAN WITH
LOTUS ROOT CAKE
59 綠茶紅豆糕
GREEN TEA WITH
RED BEAN CAKE
60 藍莓優格慕斯
BLUEBERRY
YOGURT MOUSSE
62 草莓慕斯
STRAWBERRY MOUSSE
64 柳橙慕斯
ORANGE MOUSSE
66 蘋果慕斯
APPLE MOUSSE
68 榴槤慕斯
DURIAN MOUSSE
70 巧克力慕斯
CHOCOLATE MOUSSE
72 櫻桃乳酪慕斯
CHERRY CHEESE MOUSSE
74 優酪乳慕斯
YOGURT MOUSSE
76 巧克力慕斯球
CHOCOLATE
MOUSSE BALL
78 咖啡乳酪派
COFFEE CHEESE PIE

80 水果派
FRUIT PIE
82 芋泥派
TARO PIE
84 芒果塔
MANGO TART
86 戚風蛋糕做法
HOW TO MAKE CHIFFON CAKE
88 派皮的做法
HOW TO MAKE PIE CRUST

沁涼小菜

92 玉米豆腐凍
TOFU WITH CORN JELLO
94 素肉凍
VEGETARIAN MEAT JELLO
96 翡翠豆腐
SPINACH JELLO
98 虎皮凍
TIGER SKIN JELLO
100 雞凍
CHICKEN JELLO
102 牛肉凍
BEEF JELLO
104 雞肉玉米凍
CHICKEN WITH CORN JELLO
106 蔬菜火腿凍
SPINACH WITH HAM JELLO
108 甜椒鮭魚凍
PEPPER SALMON JELLO
110 石榴海鮮凍
SEAFOOD JELLO
112 冰涼魷魚卷
COLD SQUID ROLLS

如何做出心凍美食？

讓食物凍起來的材料

要將液體的食物變成固體時，除了一般的粉類，如本食譜所使用的蓮藕粉、地瓜粉、葛粉、太白粉、馬蹄粉之外，中式烹調最常被使用的有洋菜粉、蒟蒻粉、豬皮等，而西點方面以吉利丁最爲普遍。

布丁粉

布丁粉品牌非常多，各大超市都有賣，有的已經添加了各種香料或糖份；所以使用前務必看清楚材料中水份比例及糖份含量的多寡再做適當的調整，而在本食譜裡我使用的是完全不含糖的布丁粉，其使用水份比例爲1大匙布丁粉配上6～7杯水。

如何使用布丁粉

1.將布丁粉與糖充分調勻，若未與糖充分調勻即倒入水內，布丁粉很容易浮在水面而造成結塊。

2.取適量水放入鍋內，將拌好的布丁糖粉倒入，攪拌均勻後再以小火煮至完全融化切成透明狀。

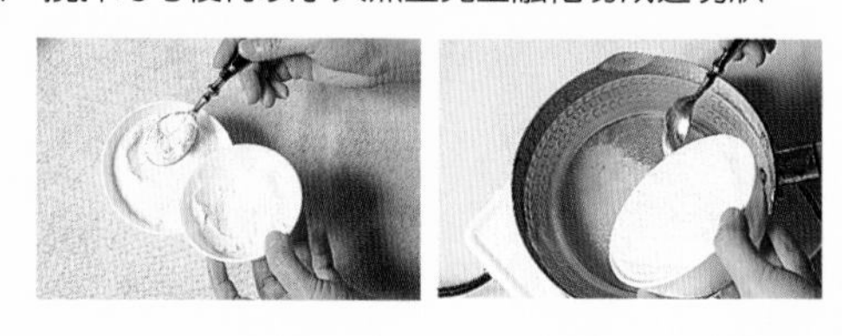

聚力T

在包裝盒上一般稱爲吉利T，其名稱由來是英文的Jelly T，我怕讀者與吉利丁Gelatin混淆，所以在本食譜中統一稱爲聚力T。

如何使用聚力T

聚力T為海藻膠的一種，屬於素食食物，使用方法與布丁粉相同，都必須先與糖混合後、放入冷水內攪拌均勻再煮，以免結塊。聚力T的凝固點很高，約50～60℃表面就開始凝固，所以不適合用來製作慕斯。使用聚力T製作時要經常攪拌不要讓表面凝結，若在製作中因凝固而無法倒入模內整型時，可再加溫使其融化再處理。

吉利丁

吉利丁名字來自英文Gelatin，有粉狀及片狀兩種，屬於動物膠，素食者不可食用，其口感較Q、凝固點較低，製作時溫度較容易控制，是製作慕斯很好的材料。

如何使用吉利丁

1.粉狀吉利丁可直接加熱水調勻，若怕會結塊，可先與糖拌勻再使用，也可直接與冷水泡發後再加熱至其融化。

2.吉利丁片必須先以冰水浸泡至軟後再放入煮熱的溶液內至其融化，在將吉利丁放入水中時，要一片一片放入，以免整疊黏在一起而泡不軟，使用冰水浸泡可使製作出的產品口感更Q。

洋菜

屬海藻膠，使用方便，是最傳統的食物凝固劑。

如何使用洋菜

只須以冷水泡軟後再放入液體內煮至融化，待涼即凝固。

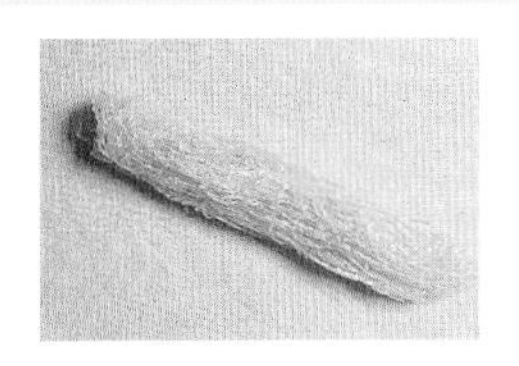

豬皮

最天然的食物凝固劑，只須與水一起熬煮後就會產生天然膠質，爲製作凍類菜餚的最好材料，其膠質對身體有很大的益處。

如何使用豬皮

先將豬皮以熱水煮約10分鐘，至內面油的部份成透明狀，取出將油徹底刮乾淨，再將表皮也刮乾淨後即可加入適量的水熬煮，為防止有腥味，通常會加入蔥、薑、酒一起煮，煮約50分鐘以上，至湯汁成乳白色即可。

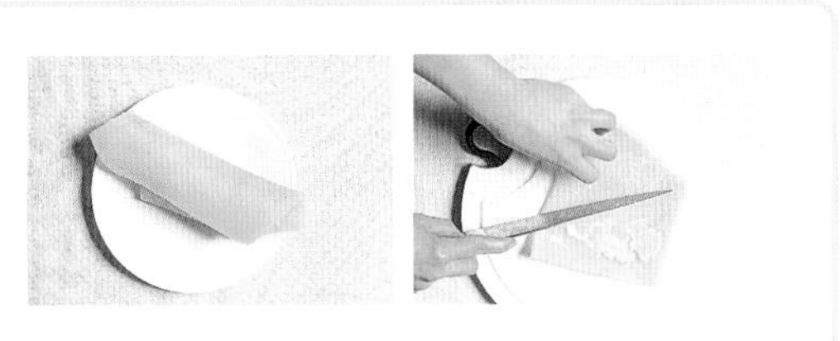

由於每一個人口感喜好不同，若你喜歡稍微硬一點，可減少些許水份，而要軟一點，則可多加水份。

其他西點材料

麵粉

有高、中、低筋三種，視配方的需要選擇不同的麵粉。

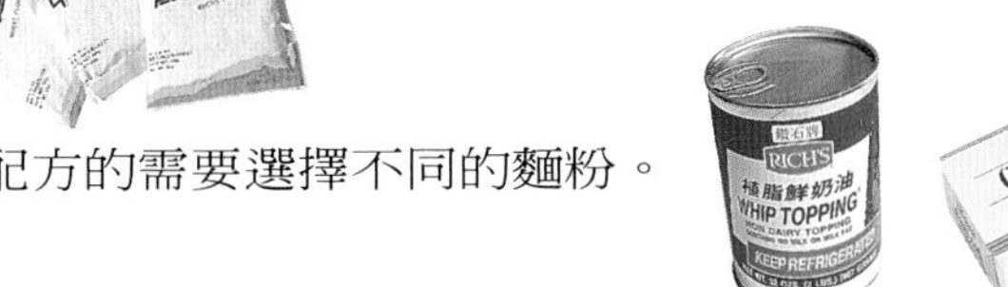

奶油

一般製作西點大多使用無鹽奶油，在超級市場都可購買得到整條長方形的奶油，使用前先放在室溫下自然軟化。

鮮奶油

有動物性及植物性兩種，未打發前的鮮奶油爲液體狀，動物性不含糖份，植物性則含有糖份。

奶油乳酪

屬於軟質乳酪的一種，是烹調乳酪蛋糕必備的材料，選購時要注意看標示爲Cream Cheese才是，有些學生會誤以爲是片狀乳酪。

How to Use Thickeners to Make Chilled Dish?

Except for the powders made from such items as lotus root, yam, arrowroot, kusu and water chestunt that can make liquid thicken and set in this cookbook, In Chinese cooking the most common thickeners are agar-agar, konnyaku powder, and pork skin. In western desserts, gelatin is the best-known.

PUDDING POWER

There are many brands of pudding powder available in any supermarket. Because some brands have flavorings or sugar added, it is necessary to read the instructions carefully to check the proportion of water and sugar in order to decide how to adjust the amount of pudding powder to be added. In this cookbook, pudding powder is assumed to contain no sugar, and the proportion of water is 1T. pudding powder to 6-7 cups of water.

 HOW TO USE PUDDING POWDER:

1. Be sure the pudding powder and sugar are well mixed, or the pudding powder will float on the surface of the water and form little clumps after the water is added.
2. After placing the proper amount of water in pan, pour in the pudding powder and sugar mixture, stir until evenly mixed, cook over low heat until sugar dissolves completely and the mixture becomes transparent.

JELLY T

Jelly T is obtained from seaweed.

 HOW TO USE JELLY T:

Jelly T is made from the seaweed and would be fine for vegetarians. The instuuctions of using jelly T are the same with pudding powder. Mix jelly T with sugar first ,cook in cold water and stir well to prevent from clumping. Jelly T gels very easily. It gels at 50-60ºC so that you have to stir constantly to prevent from setting in pan. If this occurs, simply reheat and melt. Because it is not so convenient for people to use that it would not fit to make a mousse.

GELATIN

Gelatin is obtained from animal sources by boiling hooves, bones and other parts. It is usually made into powder or large lump. Gelatin is chewier and gels at an even lower temperature. Easier to handle than jelly T so that it is an excellent ingredient for making mousse.

 HOW TO USE GELATIN:

1. Mix gelatin powder with hot water well. To keep from clumping, mix well with sugar first, or soak in cold water until swollen, then heat until dissolved.
2. Soak gelatin slices in ice water until soft, add to hot liquid, and heat until dissolved. Soak gelatin piece by piece, or it will be overlapped with each other and will not be soft.Soak in ice water makes it much chewier.

AGAR-AGAR

Agar-agar is obtained from seaweed. It is convenient and easy to use and is the most traditional thickener.

HOW TO USE AGAR-AGAR:

Just soak in cold water until soft, add to liquid, cook until dissolved, let cool and set.

PORK SKIN

Pork skin is the most natural thickener. Just cook in water and it will release its natural gelatin. It is the best ingredient for making jello as it is good for the human body.

HOW TO USE PORK SKIN:

Cook pork skin in hot water for about 10 minutes until the fat inside appears transparent, remove, scrape off the fat inside and scrape surface thoroughly, cook the skin in a suitable amount of water add scallions, ginger and wine to remove the unpleasant odor, then cook for at last 50 minutes until liquid is creamy white. If you prefer the jello eaten harder, decrease a little bit of water while cooking pork skin. Or you would like it softer, add more water to cook would be fine.

FLOUR

There are three kinds of flour, bread flour, all-purpose flour and cake flour, select as needed according to the recipes.

BUTTER

Unsalted butter is mostly used in making cake, it can be purchased in the supermarket. Soften butter at room temperature before using.

FRESH CREAM

There are two kinds of fresh cream, whipping cream (no sugar) and artificial whipping cream (with sugar). Unwhipped cream is liquid.

CREAM CHEESE

Cream cheese is a soft cheese and is also indispensable for making cheesecakes. Read label carefully when purchasing so as not to mistake it for sliced package cheeses.

MANGO TART

CHOCOLATE MOUSSE BALL

冰涼甜點
DESSERT

ED BEAN CAKE SAGO PUDDING COCONUT CAKE OSMANTHUS WITH LOTUS SEED JELLO ALMOND TOFU CAKE COKE SHERBE

ROSELLA WITH CRATAEGUS JELLO ROSELLA WITH CRATAEGUS JELLO ROSELLA WITH CRATAEGUS JELLO

洛神仙楂凍

Rosella with Crataegus Jello

INGREDIENTS:

5g. dried Rosella,
10g. Crataegus, 150g. sugar,
2T. jelly T, 600g. water

METHODS:

1 Bring water to boil in pan, add Rosella and Crataegus, cook over low heat for about 5 minutes (fig.1).

2 Remove, sieve, discard dregs, let cool, and retain the liquid.

3 Combine sugar and jelly T well, add liquid from step 2 (fig.2), return to heat, cook over low heat until completely dissolved.

4 Pour liquid into a prepared mold (fig.3), let cool, remove to refrigerator and chill until set and cold, remove, flip over onto serving plate. Serve.

材料

洛神花5公克、仙楂10公克、糖150公克、
聚力T2大匙、水600公克

做法

1 水放入鍋內燒熱後，放入洛神及仙楂，以小火煮約5分鐘（圖1）。

2 煮好後過濾、待涼。

3 糖與聚力T拌勻，倒入煮好的仙楂液（圖2），繼續以小火煮至完全融化。

4 取一模子，將煮好的液體倒入模內（圖3），待涼後放入冰箱冰至凝固且完全涼後，扣出即可食用。

NOTES: 你也可以換成→吉利丁粉 換成吉利丁粉約需使用2 1/2大匙
You can substitute 2 1/2T. gelatin powder for 2T. jelly T.

MIXED FRUIT JELLO MIXED FRUIT JELLO MIXED FRUIT JELLO MIXED FRUIT JELLO MIXED FRUIT JELLO MIXED FRUIT JELLO MIXED FRUIT

什錦水果凍

Mixed Fruit Jello

INGREDIENTS:

1 kiwi, 1/2 apple,
about 5 strawberries, 120g. sugar,
2T. jelly T, 600g. water

METHODS:

1. Mix jelly T and sugar well (fig.1); dice kiwi, apple and strawberries.
2. Combine sugar with water in pan (fig.2).
3. Cook sugar syrup over low heat until dissolved completely and transparent.
4. Place all diced fruit in mold, pour in liquid (fig.3), let cool, then chill in refrigerator for about 2 hours until set thoroughly and cold. Flip over onto serving plate and serve.

材料

奇異果1個、蘋果1/2個、草莓約5個、糖120公克、聚力T2大匙、水600公克

做法

1. 聚力T及糖調勻（圖1），將奇異果、蘋果及草莓切丁。
2. 鍋內放入水，加入調好的糖（圖2）。
3. 將糖水以小火煮至完全融化、呈透明狀。
4. 模內放入切好的水果丁，再倒入煮好的果凍液（圖3），待涼後放入冰箱，冰至完全凝固，涼透後（約需2小時），扣出即可食用。

NOTES: 你也可以換成→吉利丁粉
換成吉利丁粉約需使用2 1/2大匙。
You can substitute 2 1/2T. gelatin powder for 2T. jelly T.

YOGURT FRUIT JELLO YOGURT FRUIT JELLO YOGURT FRUIT JELLO YOGURT FRUIT JELLO YOGURT FRUIT JELLO

五彩優格果凍

Yogurt Fruit Jello

材料

柳橙汁1/2杯、葡萄汁1/2杯、葡萄柚汁1/2杯、優格約200公克、糖約120公克、聚力T2大匙、水約300公克

做法

1 糖與聚力T調勻（圖1），倒入水內，以小火煮至完全融化、呈透明狀（圖1）。

2 將煮好的溶液分成3等份，每一等份分別加入不同的果汁（圖2），快速攪拌均勻。

3 取一平面模子，鋪上保鮮膜，倒入溶液（圖3），待涼後放入冰箱冰至涼透。

4 冰好的果凍切小丁，依序放入杯內，每種果凍之間放入薄薄一層打散的優格。

＊因聚力T凝固點較高，很快就會凝結；若因中途製作太慢而造成溶液凝固無法倒入模內整型，則可放回爐上加溫，讓其再度融化。

NOTES: 你也可以換成→吉利丁粉
換成吉利丁粉約需使用2 1/2大匙。
You can substitute 2 1/2T. gelatin powder for 2T. jelly T.

INGREDIENTS:

1/2C. orange juice, 1/2C. grape juice, 1/2 grapefruit juice,
about 200g. yogurt, about 120g. sugar, 2T. jelly T, about 300g. water

METHODS:

1 Combine sugar and jelly T well (fig.1), pour into water, cook over low heat until sugar dissolves completely and is transparent (fig.1).

2 Divide liquid into 3 portions, add juice to liquid (fig.2), stir rapidly until evenly mixed.

3 Prepare a long, flat mold lined with saran wrap, pour in liquid mixture from step 2 (fig.3), let cool, then chill in refrigerator until cold.

4 Dice the jello, remove to glass in order, spread beaten yogurt between each kind of fruit jello.

＊jelly T gets set very quickly, if liquid is set before done, reheat to dissolve.

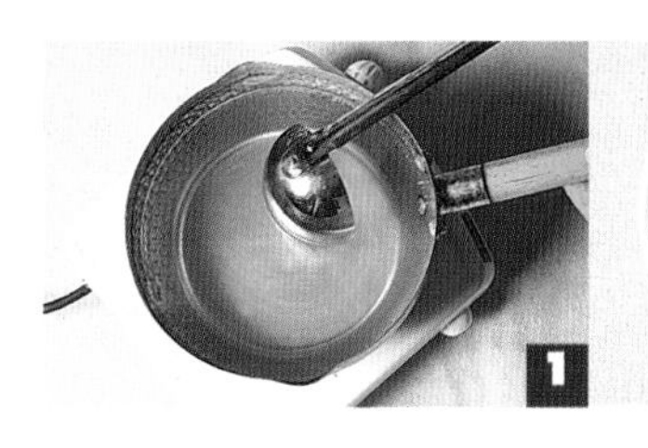
1

2

3

番茄果凍

Tomato Jello

INGREDIENTS:

2 large red tomatoes,
200g. tomato juice, 60g. sugar,
1T. jelly T 2T. honey,

METHODS:

1 Rinse tomatoes, cut horizontally about 0.5cm deep from top, remove seeds and flesh to form shell, retain tomato shells, and dice flesh.

2 Mix jelly T well with sugar, add tomato juice (fig.1), cook over low heat, stir continuously without stopping until jelly T dissolves completely.

3 Add diced tomato to step 2 (fig.2), then add honey and mix well.

4 Remove tomato juice and honey mixture to tomato shells (fig.3), let cool first, then chill in refrigerator until set and cold, remove and slice. Serve.

材料

大型紅番茄2個、番茄汁200公克、糖60公克、聚力T1大匙、蜂蜜2大匙

做法

1 紅番茄洗淨，由蒂部往下約0.5公分處橫切開來，將內部籽挖除，果肉切小丁。

2 聚力T調入糖拌勻，加入番茄汁（圖1），以小火煮至聚力T完全融化，煮時要不停攪拌。

3 番茄汁煮好後，加入果肉（圖2），再加入蜂蜜攪拌均勻。

4 將蜂蜜番茄汁盛入番茄內（圖3），待稍涼，放入冰箱冷藏，冰至凝固且涼透，取出切片即可食用。

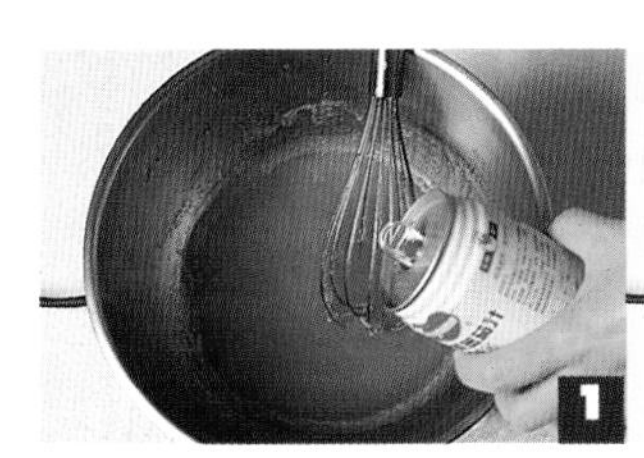

1

2

3

NOTES: 你也可以換成→吉利丁粉

換成吉利丁粉約需使用1大匙，番茄汁要減為150公克。
You can substitute 1T. gelatin powder for 1T. jelly T. If you use 1T. gelatin powder, the tomato juice should be decreased to 150g.

綠茶亞答子凍

Green Tea with Yatatz Jello

材料

綠茶粉1茶匙、聚力T2大匙、糖150公克、水350公克、椰漿1杯、亞答子約1杯

做法

1 綠茶粉與聚力T、糖攪拌均勻（圖1）。

2 鍋內放入水，再加入調好的椰漿（圖2），攪拌均勻。

3 將拌好的綠茶糖粉倒入（圖3），快速攪拌均勻，以小火煮至完全融化。

4 取1模型，放上1粒亞答子（圖4），再將煮好的綠茶溶液倒入模型內，待涼放入冰箱冷藏至完全冰透（約需2小時），扣出即可食用。

＊綠茶粉也就是一般材料行所謂的抹茶粉，使用時要先與糖拌勻，否則很容易結塊。

＊亞答子為罐頭食品，在一般超市或南洋食品材料行均可購得。

NOTES: 你也可以換成→吉利丁粉
換成吉利丁粉約需使用2 1/2大匙。
You can substitute 2 1/2T. gelatin powder for 2T. jelly T.

INGREDIENTS:

1t. ground green tea, 2T. jelly T, 150g. sugar,350g. water, 1C. coconut milk, about 1C. yatatz (vietnamese tropical fruit)

METHODS:

1 Combine green tea, jelly T and sugar well (fig.1).

2 Fill pan with water, add well-mixed coconut milk (fig.2), stir to mix well.

3 Add step 1 to step 2 (fig.3), stir rapidly until evenly mixed, cook over low heat until dissolved completely.

4 Prepare a mold, place a yatatz in center, pour in green tea mixture, let cool, chill in refrigerator for about 2 hours until thoroughly chilled, remove, flip over onto plate. Serve.

＊Ground green tea can be purchased at any tea shop, it has to be mixed well with sugar before use or it will crumble easily.

＊Yatatz is a canned fruit; you can buy it in the supermarket or the shops especially sell the Southeast Asian food and ingredients.

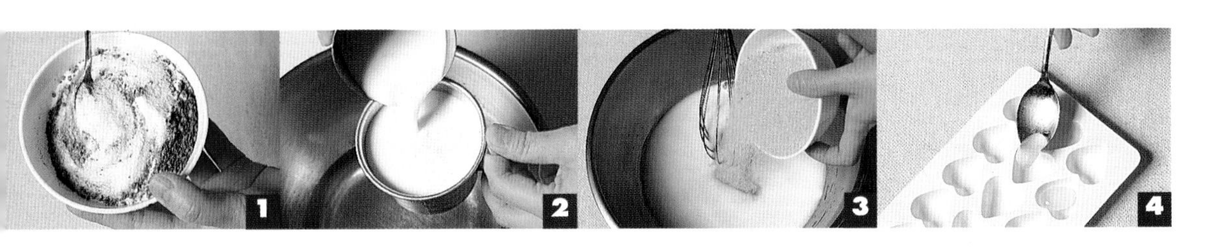

蜜汁胡蘿蔔凍

Carrot with HONEY JELLO

INGREDIENTS:

1 carrot (about 200g), 100g. sugar, 2T. jelly T powder, 300c.c. water, honey as needed

METHODS:

1. Pare carrot, cut into large chunks, place in blender, add water (fig.1), blend until mashed.
2. Mix jelly T powder and sugar together well (fig.2).
3. Place carrot juice in pan, add sugar (fig.3), and stir well. Cook over low heat until jelly T dissolves and carrot releases its flavor. Let cool, pour into molds, and refrigerate until set and cold. Flip onto plate, drip with honey and serve.

材料

胡蘿蔔1根（約200公克）、糖100公克、吉利丁粉2大匙、水300c.c.、蜂蜜適量

做法

1. 胡蘿蔔去皮切滾刀塊，加水放入果汁機（圖1），攪打成泥。
2. 吉利丁粉與糖充分攪拌均勻（圖2）。
3. 胡蘿蔔汁放入鍋內，加入吉利丁糖粉（圖3）攪拌均勻，以小火煮至吉利丁完全融化、且胡蘿蔔香味溢出後，放涼，再倒入模子內，放入冰箱冷藏至完全凝固且涼透，扣出後淋上蜂蜜即可食用。

1

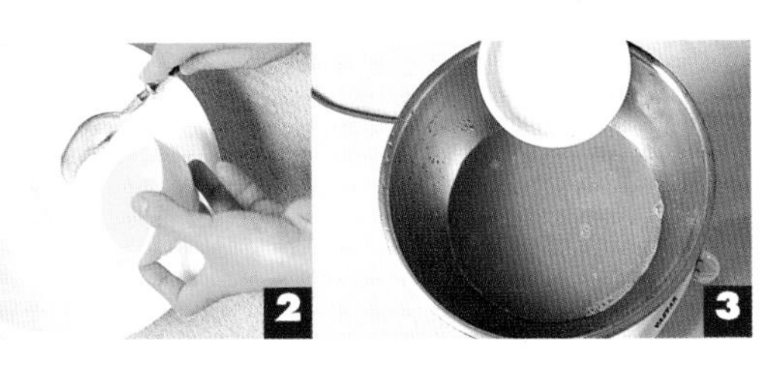

2

3

NOTES: 你也可以換成→吉利丁片or聚力T
換成吉利丁片約需使用7 1/2片，換成聚力T則約需1 1/2大匙。
You can substitute 7 1/2 slices of gelatin or 1 1/2T. jelly T for gelatin powder.

南瓜布丁

Pumpkin Pudding

INGREDIENTS:

300g. peeled pumpkin, 150g. sugar, 2T. jelly T, 500g. water

METHODS:

1. Cut pumpkin into small chunks, remove to bowl, add 2T. water, cover with saran wrap (fig.1), cook in microwave for about 5 minutes (or steam in steamer).
2. Mash pumpkin with electric mixer (fig.2), or mash with a spoon.
3. Stir pumpkin mash with 500g. water well. Combine jelly T with sugar well, add to pumpkin mash(fig.3), and stir evenly mixed. Cook slowly over low heat until jelly T dissolves completely, pour into molds, cool, then remove to refrigerator and chill for about 2 hours until cold, remove, flip onto serving plate. Serve.

材料

去皮南瓜300公克、糖150公克、
聚力T粉2大匙、水500公克

做法

1. 南瓜切小塊，放入碗內，加入2大匙水，蓋上保鮮膜（圖1），以微波爐煮約5分鐘（也可用蒸籠蒸，約需15分鐘）。
2. 將蒸好的南瓜以攪拌器打爛成泥（圖2），或以湯匙壓爛。
3. 南瓜泥加入水攪拌均勻，聚力T粉與糖調勻，倒入南瓜泥內（圖3），攪拌均勻後放在爐上以小火慢慢煮至聚力T粉完全融化，再倒入模內；待涼後放入冰箱冷藏。至完全涼透後（約需2小時），扣出即可食用。

NOTES: 你也可以換成→吉利丁粉　換成吉利丁粉約需使用2 1/2大匙。
You can substitute 2 1/2T. gelatin powder for 2T. jelly T.

雞蛋布丁

Egg Pudding

材料

蛋黃6個、水6杯、調製奶水400公克、
砂糖180公克、布丁粉1大匙、
香草精1/2茶匙、可可粉1大匙

做法

1 鍋內放入水，砂糖與布丁粉調勻拌入（圖1），攪拌均勻；以小火煮至布丁粉完全融化，取1杯調入可可粉拌勻盛入布丁杯中。

2 蛋黃加入調製奶水（圖2），攪拌均勻。

3 將拌好的蛋黃奶水過濾，放入煮好的布丁糖水內（圖3），再加入香草精，攪拌至完全混合。

4 蛋液盛入布丁模內（圖4），待涼後放入冰箱冷藏至凝固且冰透即可食用。

NOTES: 布丁粉由於品牌不同，所使用的水份量也會不同，使用前請詳細參考說明書。
Pudding powders vary from each other and the amount of water needed to add would be different. Read the instructions carefully before using.

INGREDIENTS:

6 egg yolks, 6C. water,
400g. evaporated milk,
180g. granulated sugar
1T. pudding powder,
1/2t. vanilla essence,
1T. cocoa powder

METHODS:

1 Fill pan with water, mix sugar with pudding powder well (fig.1). Cook over low heat until pudding powder dissolves completely, get 1C. pudding syrup, mix well with cocoa powder, and pour in pudding molds.

2 Add yolks to evaporated milk (fig.2), and stir well.

3 Sift step 2, add to pudding syrup from step 1 (fig.3), add vanilla essence, stir continuously until well-mixed.

4 Pour liquid in pudding molds (fig.4). Let cool, refrigerate until set and cold. Serve.

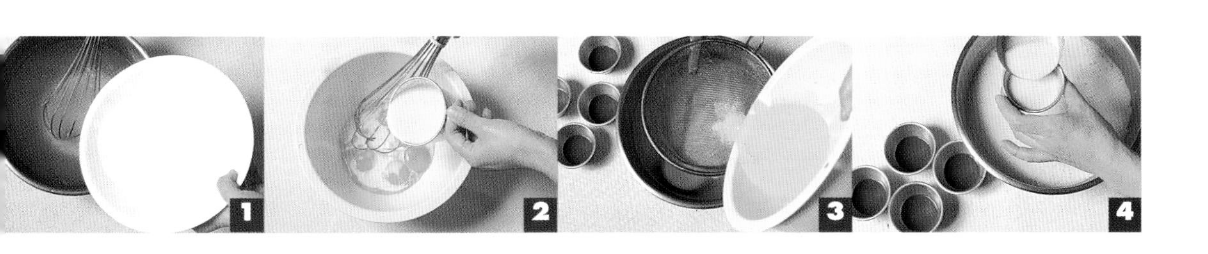

草莓涼糕

Cold Strawberry Cake

材料

草莓粉約15公克、冰水1杯、
吉利丁粉$1^1/_2$大匙、熱水1/4杯、蜂蜜3大匙、
蛋白2個、糖30公克、打好鮮奶油1杯、
椰子粉1/2杯

做法

1 草莓粉以冰水調勻（圖1）。

2 吉利丁粉以熱水調勻後倒入草莓糊內（圖2），再加入蜂蜜調勻。

3 蛋白打至起泡，加入糖，打至挺立後拌入草莓糊內（圖3），再拌入鮮奶油。

4 取1平盤，撒上椰子粉，倒入調好的草莓蛋糊，表面再撒椰子粉，放入冰箱冷藏至完全凝固且涼透，取出切小塊即可食用。

NOTES: 你也可以換成→吉利丁片
換成吉利丁片約需使用$7^1/_2$片，本食譜不適合使用其他粉類。
You can substitute $7^1/_2$ slices of gelatin for $1^1/_2$T. gelatin powder.
Other materials such as jelly T, pudding powder and agar-agar are not fitted in the recipe.

INGREDIENTS:

about15g. strawberry powder,
1C. iced water, $1^1/_2$T. gelatin powder,
1/4C. hot water, 3T. honey,
2 egg whites, 30g. sugar,
1C. whipped cream,
1/2C. shredded coconut flakes

METHODS:

1 Mix strawberry powder with iced water well (fig.1).

2 Mix gelatin powder with hot water well, pour into strawberry batter (fig.2), add honey and stir to mix well.

3 Beat egg whites until fluffy, add sugar, beat until mixture forms stiff peaks, fold into strawberry batter (fig.3), followed by whipped cream.

4 Sprinkle shredded coconut flakes over a large, flat pan, pour in strawberry batter. Sprinkle shredded coconut on surface, refrigerate until set and cold, remove and cut into pieces. Serve.

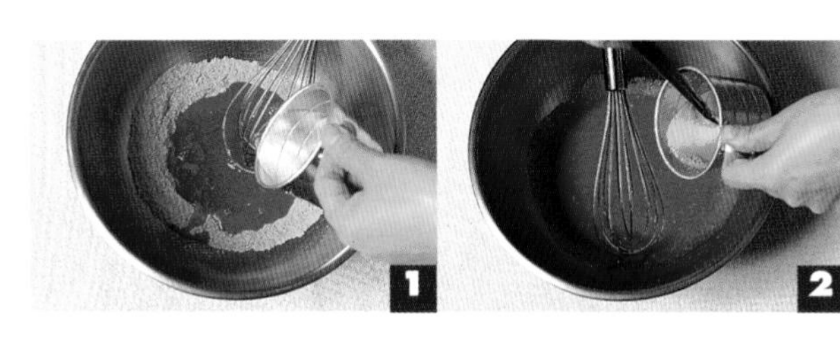
1 2

3

綠豆涼糕

Cold Mung Bean Cake

材料

去皮綠豆300公克、糖120公克、
聚力T2大匙、水3杯

做法

1 綠豆以水浸泡約6小時，取出濾乾水分。

2 將綠豆放入蒸籠內，以大火蒸約30分鐘，取出加入水，放入果汁機內攪打成泥（圖1）。

3 糖與聚力T攪拌均勻後倒入豆泥內（圖2），放爐上以小火煮至完全融化。

4 取1平盤，倒入煮好的豆泥（圖3），待涼後放入冰箱冷藏至完全凝固且涼透，約需2小時，食用時切小塊即可。

INGREDIENTS:

300g. shelled mung bean,
120g. sugar, 2T. jelly T, 3C. water

METHODS:

1 Soak beans in water for about 6 hours, remove and drain.

2 Steam beans in steamer on high heat for about 30 minutes. Remove, add 3C.water, and mash in blender with water added (fig.1).

3 Mix sugar and jelly T well, add to bean mash (fig.2), cook over low heat until dissolved completely.

4 Spread beans evenly in a large, flat pan (fig.3). Let cool and refrigerate for about 2 hours until set and cold. cup into pieces and serve.

NOTES: 你也可以換成→吉利丁粉or吉利丁片
換成吉利丁粉約需使用2 1/2大匙。
換成吉利丁片約需使用12片。
You can substitute 2 1/2T. gelatin powder or 12 slices of gelatin for 2T.jelly T.

紅豆軟糕

Soft Red Bean Cake

材料

鮮奶1杯、椰漿1杯、糖120公克、聚力T2 1/2大匙、水1杯、煮熟紅豆約300公克

做法

1 鮮奶、椰漿和水混和放入盆內，加入調勻的糖與聚力T（圖1），攪拌均勻後以小火煮至聚力T完全融化。

2 拌入紅豆攪拌均勻（圖2），待稍涼。

3 取一心形慕斯模，底部鋪上錫箔紙，再將拌好的紅豆倒入（圖3），待涼，放入冰箱冷藏至凝固且完全涼透，即可將模子取下。

INGREDIENTS:

1C. milk, 1C. coconut milk, 120g. sugar, 2 1/2T. jelly T, 1C. water, 300g. cooked red beans

METHODS:

1 Combine milk, coconut milk and water in bowl, mix sugar and jelly T well (fig.1) and cook over low heat until jelly T dissolves completely.

2 Add red beans, mix evenly (fig.2), and let cool.

3 Line bottom of heart-shaped mousse pan with aluminum foil, pour in step 2 (fig.3). Let cool , refrigerate until set and cold, and unmold. Serve.

紅豆煮法

HOW TO COOK RED BEANS

將紅豆洗淨，以水浸泡約6小時，水的高度必須是紅豆煮好後仍能完全淹蓋紅豆才可，煮沸後，改小火煮約10分鐘，熄火燜約1小時，再煮5分鐘，將水濾乾，加適量糖調味即可。

Rinse red beans and soak in water for about 6 hours (be sure water would cover beans after cooking). Bring to boil first, reduce heat to low, and cook for another 10 minutes. Remove from heat, simmer for 1 hour, then return to heat and cook for 5 minutes. Drain, add sugar to taste.

NOTES: 你也可以換成→吉利丁粉換成吉利丁粉約需使用3大匙。

You can substitute 3T. gelatin powder for 2 1/2T. jelly T.

西米布丁

Sago Pudding

材料

鮮奶2杯、椰漿1杯、西谷米150公克、糖120公克、奶油40公克、吉利丁粉$1^1/_2$大匙

做法

1 鮮奶放入盆內（圖1），加入椰漿（圖2）攪拌均勻後，以小火煮滾，放入西谷米，煮至西谷米呈透明狀。

2 吉利丁粉與糖調勻後，倒入西谷米中（圖3），攪拌至吉利丁及糖都完全溶化，再加入奶油拌勻。

3 待西谷米涼後盛入模子內，放入冰箱冷藏至凝固且涼透即可食用。

INGREDIENTS:

2C. milk, 1C. coconut milk, 150g. sago, 120g. sugar, 40g. butter, $1^1/_2$T. gelatin powder

METHODS:

1 Pour milk in bowl (fig.1), add coconut milk (fig.2), stir well, cook over low heat until boiled. Add sagos, then cook until sagos are transparent.

2 Combine gelatin powder and sugar well, add to sagos (fig.3), stir until gelatin and sugar dissolve, add butter, mix evenly.

3 Cool first, then pour into pudding molds, refrigerate until set and cold. Serve.

NOTES: 你也可以換成→聚力T
換成聚力T則約需1大匙。
You can substitute 1T. jelly T for $1^1/_2$T. gelatin powder.

COCONUT CAKE COCONUT CAKE COCONUT CAKE

椰汁雪花糕

Coconut Cake

材料

椰漿1杯、水1杯、蛋白2個、糖150公克、吉利丁片10片、椰子粉1/2杯

做法

1. 吉利丁片以冰水浸泡約10分鐘。
2. 椰漿加入水及120公克糖，以小火煮至滾，至糖完全融化後，熄火。
3. 吉利丁片捏乾水份後，放入煮好的糖液內（圖1），攪拌至完全融化，待涼備用。
4. 蛋白打至起泡後，加入剩餘的30公克糖，打至挺立（圖2）。
5. 打好的蛋白拌入已涼的椰漿糖液中（圖3），至變濃稠後，倒入鋪上保鮮膜的平盤內，上撒椰子粉後放入冰箱冷藏至凝固且冰透（約需2小時），取出切成長方形小塊即可食用。

＊若想讓椰漿糖液快速變涼，可在底部放置一盆冰水，以隔水法不停攪拌，使其變涼變稠。

＊In order to cool sugar and coconut mixture as rapidly as possible, place a basin of ice water underneath while stirring mixture continuously to cool and thicken.

INGREDIENTS:

1C. coconut milk, 1C. water,
2 egg whites, 150g. sugar,
10 slices gelatin,
1/2C. shredded coconut flakes

METHODS:

1. Soak gelatin in ice water for about 10 minutes.
2. Add water and 120g. sugar in coconut milk, cook over low heat until boiled and sugar dissolves completely, then remove from heat.
3. Remove gelatin, squeeze out excess water, add to cooked sugar syrup (fig.1), stir constantly until dissolved, let cool and set aside.
4. Beat egg whites until fluffy, add rest of 30g. sugar, beat until it forms stiff peaks (fig.2).
5. Fold egg whites into cool sugar and coconut syrup (fig.3) in bowl, stir until sticky, fill in cake pan lined with saran wrap, sprinkle with shredded coconut flakes, chill in refrigerator for about 2 hours until set and cold, remove and cut into rectangular pieces, serve.

NOTES: 你也可以換成→吉利丁粉

換成吉利丁粉約需使用2大匙，本食譜不可以布丁粉或聚力T代替，因兩者的凝固點很高，無法拌入蛋白。

You can substitute 2T.gelatin powder for 10 slices of gelatin. Do not substitute pudding powder or jelly T for gelatin because both pudding powder and jelly T would get set easily so that it is hard to fold into the egg whites.

桂花蓮子凍

Osmanthus with Lotus Seed Jello

材料

蓮子1杯、糖120公克、聚力T1大匙、
水300公克、桂花醬2大匙

做法

1 蓮子洗淨，以水浸泡約2小時至軟，濾乾水分，若使用新鮮蓮子，則不用浸泡。

2 將蓮子放入鍋內，加入3杯水（圖1），新鮮蓮子煮約10分鐘，乾燥蓮子則約煮20分鐘至軟，撈起，濾乾水分。

3 另取1鍋，放入300公克水，加入糖及聚力T混和拌勻（圖2），煮至聚力T完全融化，且呈透明狀。

4 加入煮好的蓮子及桂花醬（圖3），攪拌均勻後，盛入模型中，待涼後放進冰箱冷藏至完全凝固且冰透即可食用。

NOTES: 你也可以換成→吉利丁粉

換成吉利丁粉約需1大匙，水須減為250公克。
You can substitute 1T.gelatin powder for 1T. jelly T. If you use the gelatin powder, the water should be decreased to 250g.

INGREDIENTS:

1C. lotus seed, 120g. sugar,
1T. jelly T, 300g. water,
2T. osmanthus sauce

METHODS:

1 Rinse lotus seeds, soak in water for about 2 hours until soft, remove, drain (if fresh lotus seeds are used, skip soaking).

2 Put lotus seeds in pan with 3C. water (fig.1), cook about 20 minutes until tender, or cook fresh ones for about 10 minutes, remove and drain.

3 Fill another pan with 300g. water, add sugar and jelly T, mix well (fig.2). Cook until jelly T dissolves completely and is transparent.

4 Add cooked lotus seeds and osmanthus sauce (fig.3). Stir until evenly mixed, remove to mold. Let cool, refrigerate until set and cold. Serve.

杏仁豆腐

Almond Tofu Cake

材料

布丁粉1大匙、細砂糖120公克、水4杯、鮮奶1杯、冷開水2杯、杏仁露約1大匙

【食用時配料】

糖50公克、熱水1/2杯、冷開水2杯、什錦水果罐頭約1杯、新鮮水果適量

做法

1. 鍋內放水，布丁粉與細砂糖調勻倒入（圖1），攪拌均勻後，以小火煮至完全融化呈透明狀。
2. 加入鮮奶（圖2）及冷開水，快速攪拌均勻後，再加入杏仁露（圖3）拌勻，盛入模型內，並將表面的氣泡撈掉，待涼後放入冰箱冷藏至涼透。
3. 將配料之糖與熱水調勻至融化，加入冷開水及什錦水果罐頭，並將新鮮水果切小塊，再加入切塊的杏仁豆腐，可加入數顆冰塊，冰至涼透後即可食用。

NOTES: 你也可以換成→洋菜

換成洋菜則約需使用18公克。

You can substitute 18g. agar-agar for 1T.pudding powder.

INGREDIENTS:

1T. pudding powder,
120g. caster sugar, 4C. water, 1C. milk,
2C. cold water, 1T. almond essence

【SIDE DISHES】

50g. sugar, 1/2C. hot water,
2C. cold water,
about 1C. canned mixed fruit,
fresh fruit as desired

METHODS:

1. Fill pan with water, add pudding powder and sugar (fig.1), stir well, and cook over low heat until dissolved and transparent.
2. Add milk (fig.2) and cold water, stir rapidly until well mixed, add almond essence (fig.3). Remove to mold, and skim bubbles off surface. Let cool, refrigerate until cold and set.
3. Combine sugar and hot water with ingredients of side dish, add cold water, canned fruit, fresh fruit dics and cut almond tofu cake. Add ice cubes if desired, serve.

草莓冰寶

Strawberry Sherbet

材料

草莓300公克、水1杯、糖120公克

做法

1. 草莓去蒂，洗淨後切小塊，放入果汁機內，加入1/2杯水打成泥。
2. 剩餘的水與糖混合，放入鍋內，以小火煮至糖融化。
3. 打好的草莓泥加入糖水內，煮至滾後，改小火熬煮約10分鐘，至濃稠後倒入器皿內，待涼放入冰箱冷凍庫內冰至硬。
4. 冰好取出，挖鬆後整型即可食用。

＊若因天氣太熱或動作較緩，冰寶挖鬆後很容易再軟化，可將其整型後放入冰箱再冷凍約1小時後食用。

INGREDIENTS:

300g. strawberry, 120g. sugar, 1C. water

METHODS:

1. Remove stems from strawberries and discard, rinse, cut into small pieces, blend in blender with 1/2C. water added .
2. Combine rest of the water with sugar in pan, cook over low heat until sugar dissolves.
3. Add strawberries to sugar, cook until boiled, reduce heat to low and cook for 10 minutes until thick and sticky, pour into mold, freeze in freezer until hard.
4. Remove, unmold and serve.

＊If sherbet happen to melt before serving, reshape and return to freezer and freeze again for about 1 hour before serving .

芒果冰寶

Mango Sherbet

材料

芒果肉250公克、糖120公克、水1/2杯

做法

1 芒果切小塊，加入糖和水（圖1），以小火煮約10分鐘。

2 以攪拌器將煮好的芒果打成泥，也可用果汁機打，但不要打得太細，至還有點果肉即可（圖2）。

3 待芒果涼後，放入冰箱冷凍庫內冷凍，中途取出攪拌約2、3次，至完全結冰即可。

INGREDIENTS:

250g. mango flesh, 120g. sugar, 1/2C. water

METHODS:

1 Dice mango, add sugar and water (fig.1), cook over low heat for about 10 minutes.

2 Mash mango with electric mixer or blender, leaving only a few chunks of mango(fig.2) .

3 Let cool, freeze mango mash until thoroughly frozen, remove and stir 2 or 3 times as it freezes.

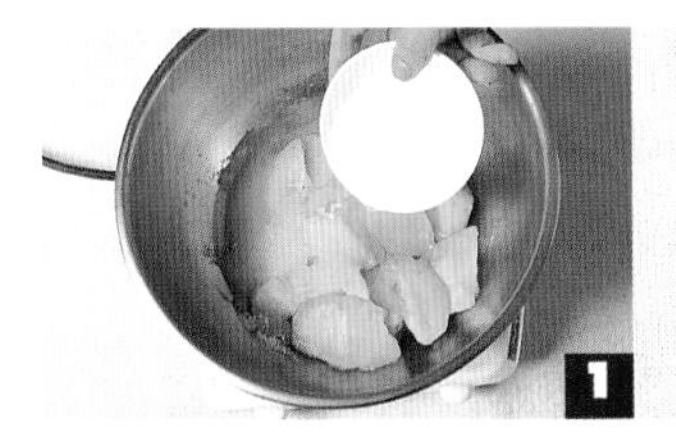

1

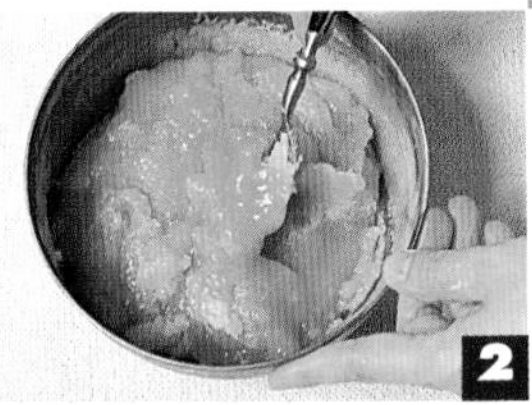

2

可樂冰寶

Coke Sherbet

INGREDIENTS:

$1\frac{1}{2}$C. coke, 40g. sugar,
1T. gelatin powder

METHODS:

1 Mix gelatin powder with sugar well (fig.1), stir to mix.

2 Pour coke in pan, add sugar (fig.2), and mix well. Cook over low heat until gelatin dissolves.

3 Let cool and refrigerate until chill and set. Stir 2 or 3 times as it chills (fig.3).

材料

可樂$1\frac{1}{2}$杯、糖40公克、吉利丁粉1大匙

做法

1 吉利丁粉放入糖內（圖1），攪拌均勻。

2 可樂放入鍋內，加入調勻的糖（圖2），攪拌均勻，以小火煮至吉利丁粉完全融化。

3 煮好後待涼，放入冰箱冷凍庫內，冰至完全凝固，中途需拿出攪拌約2、3次（圖3）。

NOTES: 你也可以換成→吉利丁片or聚力T
換成吉利丁片約需使用5片，換成聚力T則約需2茶匙。
You can substitute 5 slices of gelatin or 2t. jelly T powder for 1T. gelatin powder.

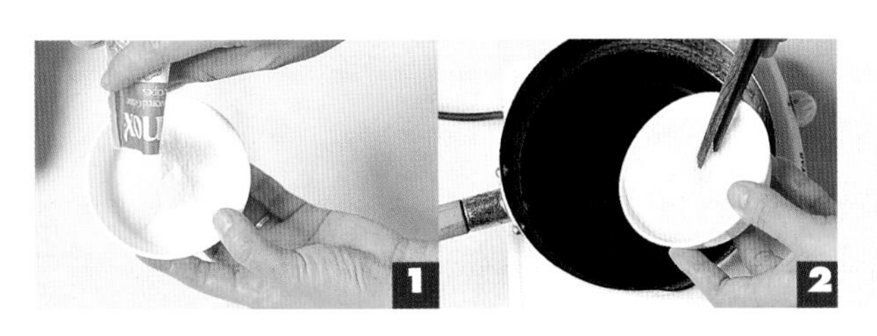

1 2

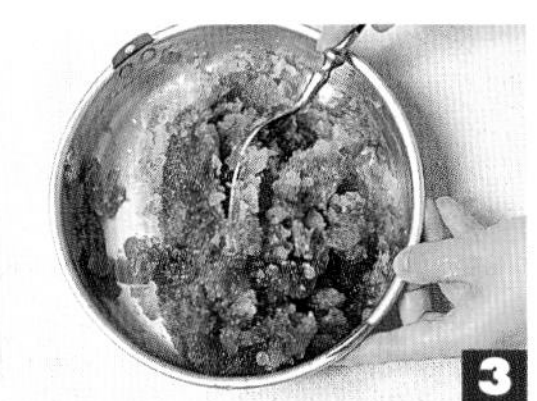

3

香草冰淇淋

Vanilla Ice Cream

材料

鮮奶3杯、鮮奶油1杯、雞蛋3個、
香草精1大匙、糖150公克、吉利丁粉1大匙

做法

1. 取100公克糖與吉利丁粉調勻。
2. 不鏽鋼盆內放入鮮奶及鮮奶油，再加入調勻吉利丁的糖（圖1）攪拌均勻，以小火慢慢煮至吉利丁完全融化。
3. 雞蛋放入盆內打至起泡（圖2），加入剩餘的50公克糖，繼續打至呈乳白色。
4. 取1杯煮好的奶水，加入蛋糊內攪拌均勻（圖3），再將所有奶水與蛋糊調勻，放在爐上以小火煮至約80ºC（火不要太大，以免煮滾）。
5. 煮好後加入香草精拌勻（圖4），以隔冰水方式攪拌至成糊狀後，放入冰箱冷凍庫內冰至凝固，取出攪拌均勻，再放入冷凍庫，冰至完全凝固。

NOTES: 你也可以換成→吉利丁片

換成吉利丁片約需使用5片，本食譜不適合使用其他粉類。

You can substitute 5 slices of gelatin for 1T.gelatin powder.
Other materials such as jelly T, pudding powder and agar-agar are not fitted in the recipe.

INGREDIENTS:

3C. milk, 1C. whipping cream, 3 eggs, 1T. vanilla essence, 150g. sugar, 1T. gelatin powder

METHODS:

1. Mix 100g. sugar with gelatin powder well.
2. Place milk and whipping cream in a stainless steel mixing bowl, add step1 (fig.1), and mix well. Cook over low heat until gelatin dissolves.
3. Beat eggs with electric mixer until fluffy (fig.2), add rest of 50g. sugar, and continue to beat until white.
4. Cook 1C. milk syrup, add to eggs, and mix evenly (fig.3). Then add step 2, cook over low heat until about 80ºC (Do not overheat to boil).
5. Add vanilla essence (fig.4), in double-boiler with ice water at bottom, stir mixture until thickened. Freeze in freezer until set. Remove, stir well, return to freezer and freeze until thoroughly frozen. Serve.

核桃巧克力冰淇淋

Walnut Chocolate Ice Cream

材料

鮮奶3杯、糖100公克、鮮奶油1杯、雞蛋3個、香草精1茶匙、吉利丁片5片、巧克力100公克、烤好核桃丁60公克

做法

1. 吉力丁片1片片放入冰水內，浸泡至軟，約10分鐘。
2. 鮮奶加入鮮奶油與50公克糖煮至糖化後，加入捏乾水分的吉力丁片，攪拌至完全融化。
3. 雞蛋打至起泡後，加入剩餘的50公克糖打至濃稠呈乳白色後，加入1杯煮好的鮮奶調勻後，再全部倒回奶水內（圖1）。
4. 繼續以小火煮至約80°C，不要煮滾（圖2），煮好後取出，加入香草精，再以隔冰水方式攪拌成糊狀。
5. 巧克力以隔熱水法融化後加入蛋糊內（圖3），攪拌均勻，盛入不鏽鋼器皿內，拌入核桃丁，放進冷凍庫內，冰至完全結冰，中途拿出攪拌2、3次。

INGREDIENTS:

3C. milk, 100g. sugar,
1C. whipping cream,
3 eggs, 1t. vanilla essence,
5 slices gelatin, 100g. chocolate,
60g. roasted walnut dice

METHODS:

1. Soak gelatin slice by slice in ice water for about 10 minutes until soft.
2. Add whipping cream and 50g. sugar to milk, cook until sugar dissolves, add draining gelatin, and stir until completely dissolved.
3. Beat eggs until fluffy, add the remaining 50g. sugar, beat until thick and white. Add 1C. cooked milk, stir well, then pour back into step 2 (fig.1).
4. Continue to cook over low heat to 80°C, do not bring to boil (fig.2), remove from the heat, and add vanilla essence. stir in double-boiler until thick.
5. Melt chocolate in double-boiler, add to step 4 (fig.3), and stir constantly to mix well. Remove to stainless steel mixing bowl, sprinkle walnut dice over on top, place in freezer and freeze until thoroughly frozen. Remove and stir 2 or 3 times as it freezes.

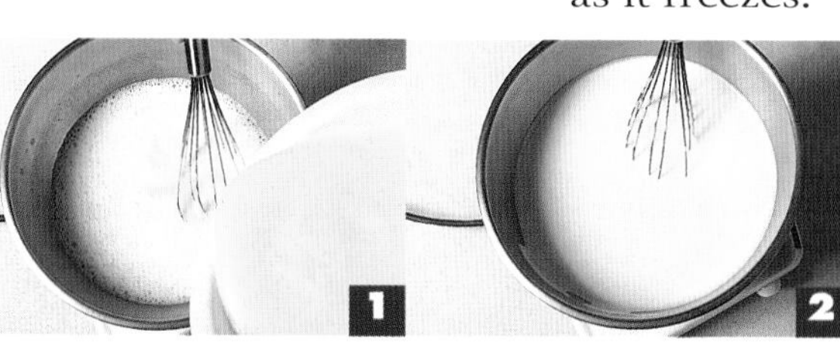

1 2

3

Soft Orange Candies Soft Orange Candies Soft Orange Candies Soft Candies Soft Orange Candies Soft Orange Candies Soft Orange

香橙軟糖

Soft Orange Candies

INGREDIENTS:

5 orange, 250g. sugar,
6T. gelatin powder, 1C. hot water,
about 2T. kusu powder

METHODS:

1 Combine sugar and gelatin powder well together (fig.1).

2 Add hot water, stir well until sugar dissolves completely.

3 Halve oranges, squeeze out about 250g. juice (fig.3).

4 Add sugar syrup to orange juice (fig.4), mix thoroughly.

5 Line mold with saran wrap, pour in orange juice mixture, let cool, and refrigerate until set.

6 Brush candy surface with a thin layer of kusu, flip over carefully onto plate, remove saran wrap, brush surface with a layer of kusu powder again, cut up into small flower candies with cutter, shake slightly to keep from sticking together.

* If gelatin powder is mixed with cold water, let stay for 5 minutes until swolled and thickened. Return to heat or heat in microwave for 1.5 minute until gelatin dissolves completely, then add juice.

材料

香吉士5個、糖250公克、吉利丁粉6大匙、熱水1杯、太白粉約2大匙

做法

1 糖與吉利丁粉調勻（圖1）。

2 加入熱水拌勻（圖2），至完全融化。

3 將香吉士切半後，搾汁，約可搾250公克果汁（圖3）。

4 果汁加入吉利丁糖液（圖4）攪拌均勻。

5 取一平面模子，鋪上保鮮膜，倒入溶液，待涼後放入冰箱冷藏至凝固。

6 將凝固的軟糖表面刷上薄薄一層太白粉，小心扣出，將保鮮膜撕去，表面再刷一層太白粉；再以小花模壓出一塊塊小小的軟糖，輕輕篩動，使其不會沾黏在一起即可。

* 若使用冷水調吉利丁粉，調好後需先放置約5分鐘，見其膨脹成濃稠，再加溫或以微波爐加熱約1分半鐘，至其完全融化後再加入果汁。

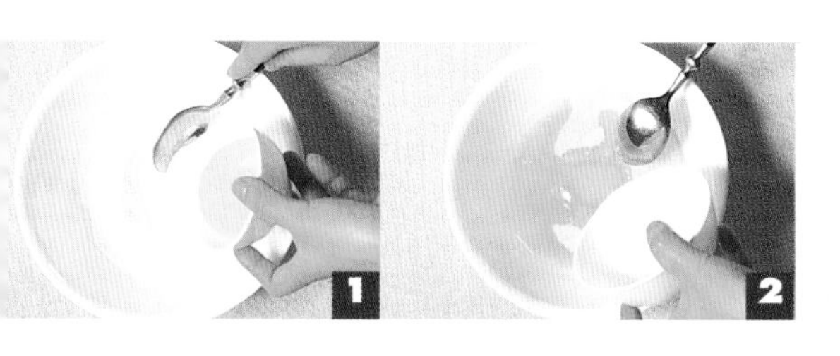
1 2

3

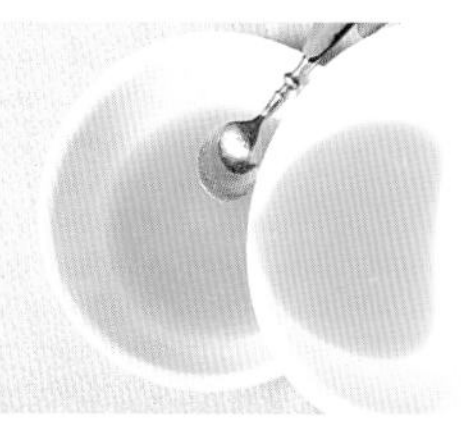
4

葡萄軟糖

Soft Grape Candies

材料

葡萄汁1杯、糖80公克、吉利丁片15片、椰子粉1杯

做法

1 吉利丁片1片片放入冰水中（圖1），浸泡約10分鐘至軟。

2 葡萄汁及糖放入鍋內，以小火煮至糖融化後熄火，吉利丁片捏乾水分後放入（圖2），攪拌至吉利丁完全融化。

3 拌好的果汁倒入四方平盤內，至涼後放入冰箱冰至凝固取出，表面撒上椰子粉（圖3），翻面再撒上椰子粉。

4 將軟糖切成四方小塊，放入調理盆內，再放入剩餘的椰子粉與軟糖拌勻（圖4），使其不沾黏在一起即可。

NOTES: 你也可以換成→吉利丁粉
換成吉利丁粉約需使用3大匙。
You can substitute 3T.gelatin powder for 15 slices of gelatin.

INGREDIENTS:

1C. grape juice, 80g. sugar,
15 slices gelatin,
1C. shredded coconut flakes

METHODS:

1 Soak gelatin in iced water one by one (fig.1) for about 10 minutes until soft.

2 Pour grape juice and sugar in pan, cook over low heat until sugardissolves, remove from heat. Remove gelatin, squeeze out excess water, add to grape juice (fig.2), and stir until gelatin dissolves completely.

3 Pour juice into a square mold. Let cool, refrigerate until set. Remove, sprinkle shredded coconut on surface (fig.3), flip over, and sprinkle shredded coconut flakes on the other side.

4 Cut square into cubes, remove to mixing bowl, add rest of shredded coconut, mix well to avoid candies sticking together.

枸杞馬蹄糕

Lycium Berry and Water Chestnut Cake

材料

馬蹄粉150公克、太白粉50公克、澄粉50公克、水5杯、糖200公克、荸薺200公克、枸杞子3大匙

做法

1. 馬蹄粉混合太白粉、澄粉，再加1杯水調勻。
2. 剩餘的水與糖煮至滾，且糖融化，沖入粉糊內（圖1），快速攪拌至呈濃稠糊狀（圖2）。
3. 枸杞子以水洗淨，荸薺去皮後切小丁；將枸杞子及荸薺拌入粉糊內（圖3）。
4. 拌勻後盛入抹油的烤模內，放入蒸籠，以大火蒸約30分鐘，至呈透明狀；中心沒有白色粉糊時即可；待涼切片食用。

＊若糖水倒入粉糊時，無法呈濃稠糊狀，可將其放回爐上，以小火煮至濃稠；但火不可太大，同時要不停攪拌，以防底部燒焦。

＊In step 2, if batter is not thick and sticky after the mixture is added, return to heat, cook over low heat until thick (do not overcook to boil); stir continuously to avoid burning.

INGREDIENTS:

150g. ground water chestnut,
50g. kusu powder,
50g. wheat starch,
5C. water, 200g. sugar,
200g. water chestnut,
3T. lycium berry

METHODS:

1. Combine water chestnut flour with kusu powder,wheat starch and 1 cup of water, stir to mix well.
2. Bring remainding water and sugar to boil, let sugar dissolve completely, pour into water chestnut mixture (fig.1), stir rapidly until mixture becomes thick and sticky (fig.2).
3. Rinse lycium berries in water; peel water chestnuts, dice finely; add lycium berries and water chestnuts to batter (fig.3).
4. Pour batter into a greased baking pan, steam in steamer on high for about 30 minutes until transparent and no white batter remains in center, let cool before slicing. Serve.

COLD TARO CAKE COLD TARO CAKE COLD TARO CAKE

芋頭涼糕

Cold Taro Cake

材料

去皮芋頭300公克、太白粉50公克、
葛粉200公克、澄粉50公克、水900公克、
糖150公克

做法

1 芋頭取200公克切薄片，放入蒸籠蒸熟，約需20分鐘；若以微波爐烹調，則將芋頭放入碗內，加1大匙水，包上保鮮膜，以強火微波約6分鐘，至熟後拌入50公克糖，再趁熱壓碎（圖1）。

2 葛粉與太白粉、澄粉混合，調入300公克水攪拌均勻。

3 將剩餘的100公克芋頭切細絲，加入600公克水，煮至滾後，改小火繼續煮約5分鐘，加入剩餘的100公克糖至糖化，趁熱沖入調好的粉糊內（圖2）至呈濃稠麵糊狀。

4 平面模子鋪上玻璃紙，取1/2麵糊鋪平，再放上芋泥（圖3），並在表面將剩餘的1/2麵糊鋪上，放入蒸籠以大火蒸約20分鐘至熟後，待涼切成長條塊狀。

INGREDIENTS:

300g. peeled taro, 50g. kusu powder,
200g. arrowroot powder,
50g. wheat starch,
900g. water, 150g. sugar

METHODS:

1 Slice 200g. of taro thinly, steam in steamer for about 20 minutes until done; or place taro in bowl with 1C. water, wrap in saran wrap and cook in microwave on high for 6 minutes until done, stir in 50g. sugar, mash while still hot (fig.1).

2 Combine arrowroot powder, kusu powder and wheat starch well, add 300g.
water, stir constantly until well-mixed.

3 Shred rest of taro, add 600g. water, bring to boil, reduce heat to low, continue to cook for another 5 minutes, add 100g. sugar, cook until sugar dissolves, pour in step 2 (fig.2), stir until thick and sticky.

4 Line a long rectangular baking mold with parchment paper, spread half of taro batter evenly over the paper, topped with shredded taro (fig.3), then cover with remaining taro batter, steam in steamer on high for about 20 minutes until done, let

綠豆蓮藕糕

Mung Bean with Lotus Root Cake

材料

蓮藕粉200公克、地瓜粉50公克、太白粉30公克、水3杯、糖150公克、蓮藕100公克、煮熟綠豆1杯

做法

1. 蓮藕粉加上地瓜粉、太白粉，再加入$1^{1}/_{2}$杯水，攪拌均勻。
2. 蓮藕洗淨切細絲，加入$1^{1}/_{2}$杯水，以小火煮約10分鐘，加入糖煮至糖化，快速沖入調勻的粉糊內（圖1），至呈濃稠粉漿狀。若無法成粉漿，可放回爐上以微火稍煮一下，煮時注意不要讓底部燒焦，也可用隔水法煮。
3. 拌入煮熟的綠豆（圖2），倒入平盤內鋪平，再以大火蒸約30分鐘，取出待涼後切塊食用。

INGREDIENTS:

200g. lotus root powder, 50g. yam flour, 30g. kusu powder, 3C. water, 150g. sugar, 100g. lotus root, 1C. cooked mung bean

METHODS:

1. Combine lotus root powder with yam flour, kusu powder, and $1^{1}/_{2}$C. water, stir constantly to mix well.
2. Rinse lotus root, shred finely, and add $1^{1}/_{2}$C. water. Cook over low heat for about 10 minutes, add suger, then cook until sugar dissolves. Pour rapidly into step 1 (fig.1), stir fast until thickened, or return to heat and cook over very low heat until thickened(Do not boil to be burnt, or use double-boiler to prevent burning).
3. Add cooked mung beans (fig.2), spread across a flat pan, steam on high for about 30 minutes. Remove, let cool, cut into piece and serve.

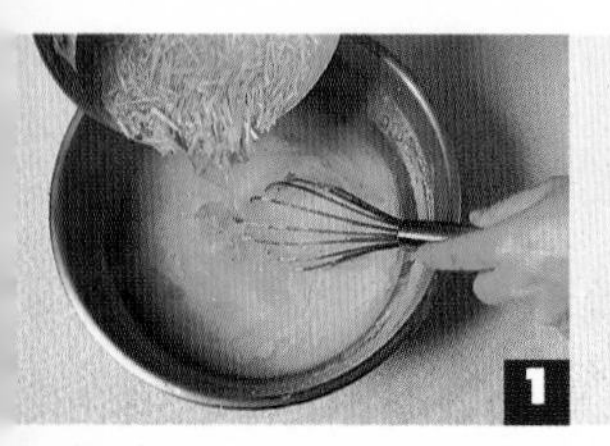

1

2

綠茶紅豆糕

Green Tea with Red Bean Cake

材料

地瓜粉120公克、太白粉30公克、綠茶粉2茶匙、鮮奶1杯、水1 1/2杯、糖100公克、煮熟紅豆約200公克（紅豆煮法請參照P.33紅豆軟糕）

做法

1 地瓜粉與太白粉、綠茶粉混和，加入鮮奶（圖1）攪拌均勻。

2 水與糖放入鍋內，以小火煮至滾，快速沖入混和好的粉糊內。

3 將其攪拌成粉漿（圖2），若無法立刻成漿，可在爐上以極小火再煮一下，煮時要不停攪拌，以免底部燒焦。

4 拌入煮熟的紅豆，整個平鋪在器皿上，放入蒸籠內，以大火蒸約30分鐘，待涼切塊。

INGREDIENTS:

120g. yam flour, 30g. kusu powder, 2t. ground green tea, 1C. milk, 1 1/2C. water, 100g. sugar, 200g. cooked red beans (See p.33 SOFT RED BEAN CAKE)

METHODS:

1 Combine yam flour, kusu powder and ground green tea, add milk (fig.1), and mix well.

2 Pour water and sugar in pan, cook over low heat until boiled, and pour rapidly into step 1.

3 Stir step2 until thick and sticky(fig2). If it is not thickened immediately, reture to cook over very low heat and stir rapidly to prevent from burning at the bottom.

4 Add cooked red beans, spread evenly in pan, remove to steamer, then steam on high for about 30 minutes. Let cool, cut into pieces and serve.

1

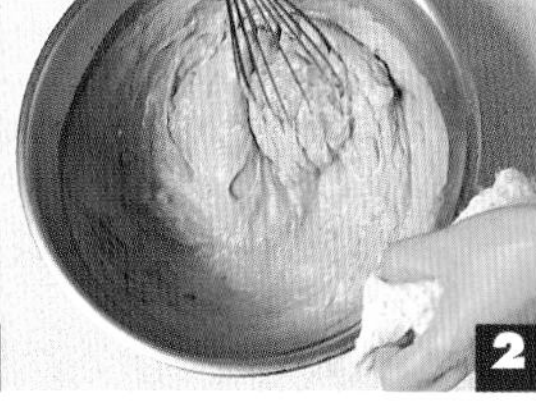

2

藍莓優格慕斯

Blueberry Yogurt Mousse

INGREDIENTS:

150g. blueberry juice,
about 50g. blueberry, 80g. sugar,
100g. yogurt, 1C. whipped cream,
6 slices gelatin

METHODS:

1 Soak gelatin in ice water slice by slice for about 10 minutes until soft.

2 Pour blueberry juice and blueberries in pan, cook over low heat with sugar until sugar dissolves. Remove gelatin, squeeze out excess water, add to juice (fig.1), and stir until gelatin dissolves.

3 Stir yogurt to pieces, add to step 2 (fig.2), and mix well.

4 Add whipped cream (fig.3), mix well, and pour into cups. Let cool, refrigerate until set and cold. Serve.

材料

藍莓汁150公克、藍莓果粒約50公克、糖80公克、優格100公克、打發鮮奶油1杯、吉利丁片6片

做法

1 吉利丁片1片片放入冰水中，浸泡約10分鐘至軟。

2 藍莓汁及果粒放入鍋內，加糖，以小火煮至糖融化，吉利丁片捏乾水分後放入（圖1），攪拌至吉利丁融化。

3 優格攪碎後拌入藍莓內（圖2），攪拌均勻。

4 再加入鮮奶油（圖3），拌勻後倒入杯中，待涼放入冰箱冷藏室冰至凝固且冰透即可。

NOTES: 你也可以換成→吉利丁粉

換成吉利丁粉約需使用1大匙，製作慕斯最好使用吉利丁，口感較好。

You can substitute 1T.gelatin powder for 6 slices of gelatin. Use gelatin to make a mousse would taste better.

1

2

3

草莓慕斯

Strawberry Mousse

材料

雞蛋2個、鮮奶1/2杯、草莓粉15公克、
吉利丁粉$1^1/_2$大匙、熱水1/4杯、糖60公克、
打發鮮奶油$1^1/_2$杯、新鮮草莓5顆、櫻桃4顆

做法

1. 吉利丁與熱水調勻備用，蛋白與蛋黃分開，分別放入兩個打蛋盆中。
2. 鮮奶加入蛋黃及30公克糖攪拌均勻，加入草莓粉拌勻（圖1），再加入吉利丁液。
3. 蛋白打至起泡後，加入剩餘的30公克糖打至挺立，拌入草莓糊中（圖2），再加入1杯打好鮮奶油，攪拌均勻。
4. 草莓去蒂洗淨後擦乾水分，切小丁，拌入草莓糊中（圖3），倒入6吋之心型慕斯模中，放入冰箱冷藏至凝固且涼透，脫模後後表面以剩餘的鮮奶及櫻桃裝飾。

NOTES: 你也可以換成→吉利丁片
換成吉利丁片約需使用8片。
You can substitute 8 slices of gelatin for $1^1/_2$T. gelatin powder.

INGREDIENTS:

2 eggs, 1/2C. milk,
$1^1/_2$T. gelatin powder, 1/4C. hot water,
60g. sugar, 15g. strawberry powder,
$1^1/_2$C. whipped cream,
5 fresh strawberries, 4 fresh cherries

METHODS:

1. Mix gelatin powder with hot water well. Separate egg whites from yolks, remove to two separate mixing bowls.
2. Add milk to yolks with 30g. sugar, combine thoroughly, add strawberry powder (fig.1). Mix well, then add gelatin water.
3. Beat whites until fluffy, add remaining 30g. sugar, beat until mixture forms stiff peaks. Fold into strawberry batter (fig.2), then add 1C. whipped cream, stir until well-blended.
4. Remove stems of strawberries, rinse, drain with paper towel, dice, add to strawberry batter (fig.3). Spread evenly in a 6-inch heart-shaped mousse pan, refrigerate until cold and set, unmold and garnish with remaining whipped cream and cherries. Serve.

1

2

3

MOUSSE ORANGE MOUSSE ORANGE MOUSSE ORANGE MOUSSE

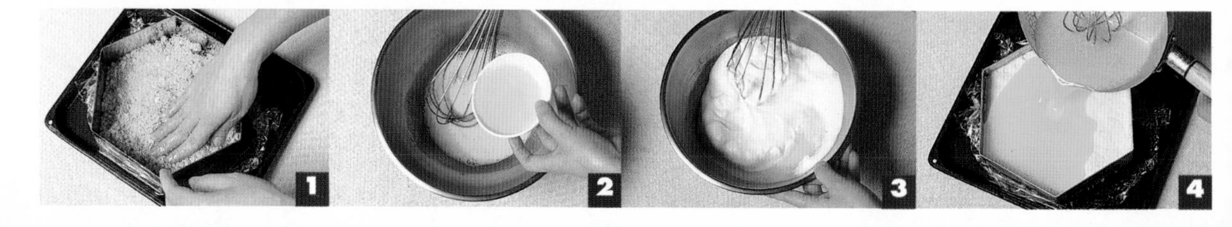
1
2
3
4

柳橙慕斯

Orange Mousse

材料

香吉士1/2個、柳橙汁3/4杯、
蘇打餅干約10片、雞蛋2個、細砂糖100公克、
吉利丁粉$2^1/_2$大匙、融化奶油60公克

做法

1 餅干壓碎，可用機器打碎，也可將餅干裝進塑膠袋中，再以擀麵棍擀碎，拌入奶油後填入模型內，盡量以手按壓至緊密（圖1）。

2 將雞蛋的蛋白蛋黃分開，蛋黃拌入1/4杯柳橙汁；取$1^1/_2$大匙吉利丁，加入1/4杯熱開水調勻，待稍涼（約50°C），拌入蛋黃內（圖2）。

3 蛋白打至起泡，加入50公克糖，打至濃稠、糖化；蛋黃液底部隔冰水攪拌至濃稠，拌入蛋白（圖3），快速攪拌均勻，再倒入模內，放進冰箱冷藏至表面凝固。

4 剩餘的1大匙吉利丁與50公克糖攪拌均勻，沖入1/2杯熱水攪拌至溶化，加入剩餘的1/2杯柳橙汁攪拌均勻，至稍涼約45°C，淋在做好的慕斯上（圖4），放入冰箱冷藏約2小時，表面以香吉士片裝飾。

＊加入吉利丁的蛋黃液，以隔冰水的方式攪拌，可使其更快濃稠，若未濃稠即拌入蛋白，容易造成蛋白與蛋黃糊分開成兩層的狀況，但也不可將蛋黃攪得太稠以致變硬，而無法拌入蛋白。

＊Stirring yolk mixture in double boiler will speed thickening. If whites are added to yolks before they are thick and sticky, they will not combine well and will separate into two layers. If yolk mixture is too thick, it will not fold into the whites.

INGREDIENTS:

1/2 orange, 3/4C. orange juice, about 10 soda crackers, 2 eggs, 100g. caster sugar, $2^1/_2$T. gelatin powder, 60g. melt butter

METHODS:

1 To make the pie shell：place crackers in plastic bag and crush with rolling pin or a blender may be used, then stir in butter well, press into pie pan.

2 Separate egg whites from yolks, add 1/4C. orange juice to yolks; add 1/4C. boiling hot water to 1/2T. gelatin, stir well, cool until 50°C, add to yolks (fig.2).

3 Beat whites until fluffy, add 50g. sugar, beat until thick and sticky and sugar dissolves; in double-boiler stir yolk mixture until thick and sticky, add whites to yolks (fig.3), stir rapidly until evenly mixed, pour into mold, chill in refrigerator until surface sets.

4 Mix remaining 1T. gelatin with 50g. sugar well, pour in 1/2C. hot water, stir until dissolved, add remaining 1/2C. orange juice, cool until 45°C, drip over mousse (fig.4), return to refrigerator and chill for about 2 hours, garnish with orange slices. Serve.

蘋果慕斯

Apple Mousse

材料

蘋果2個、鮮奶200公克、吉利丁粉2大匙、糖80公克、蛋黃3個、打發鮮奶油1杯、檸檬汁1大匙

【表面裝飾材料】

蘋果2個、糖150公克、水150公克、櫻桃1/2個

做法

1. 先做表面裝飾：蘋果2個對半切後，去除果核及蒂，再切成薄片；糖與50公克水煮成焦糖後，加入剩餘的100公克水，攪拌均勻加入蘋果片（圖1），煮至蘋果軟且入味後，待涼，整齊排列在湯碗底部（圖2）。
2. 蘋果削皮、去核去蒂，切滾刀塊，加入檸檬汁拌勻，再加入鮮奶，放入果汁機內（圖3）攪打成泥。
3. 將蘋果泥加入吉利丁粉、糖，放入鍋內，以小火煮至吉利丁融化，待稍溫，拌入蛋黃，再以隔冰水方式攪拌至涼且成稠狀，加入鮮奶油拌勻後，倒在排好的蘋果片上，放入冰箱冷藏，至凝固且完全涼透，約需2小時。

INGREDIENTS:

2 apples, 200g. milk,
2T. gelatin powder, 80g. sugar,
3 egg yolks, 1C. whipped cream,
1T. lemon juice

【GARNISH】

2 apples, 150g. sugar, 150g. water,
1/2 cherry

METHODS:

1. To make the garnish：halve apples, remove cores and stems, slice thinly. Cook sugar and 50g. water into caramel, add remaining 100g. water, stir well. Add apple slices (fig.1), cook until soft and absorbed. Let cool, line in order at bottom of bowl (fig.2).
2. Pare apples, remove cores and stems, cut into pieces, add lemon juice, and mix well. Add milk, remove to blender (fig.3), and blend until well mashed.
3. Add gelatin powder and sugar to apple mash, cook in pan over low heat until gelatin dissolves. Wait until temperature drops to warm, add yolks. In a double-boiler with ice water at bottom, stir until cool and thickened. Add whipped cream, mix well, pour over apple slices, and refrigerate for about 2 hours until set and cold. Serve.

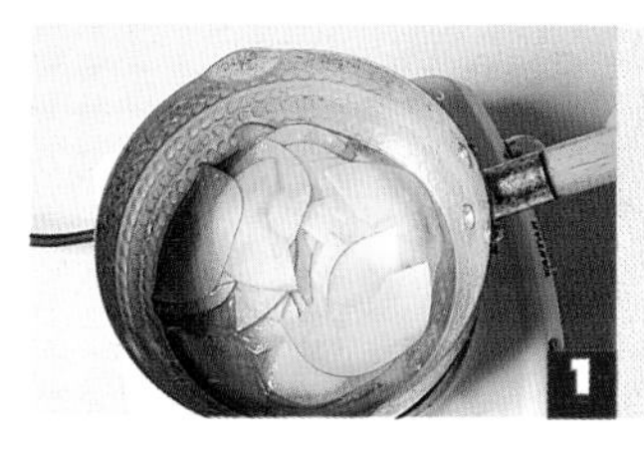
1

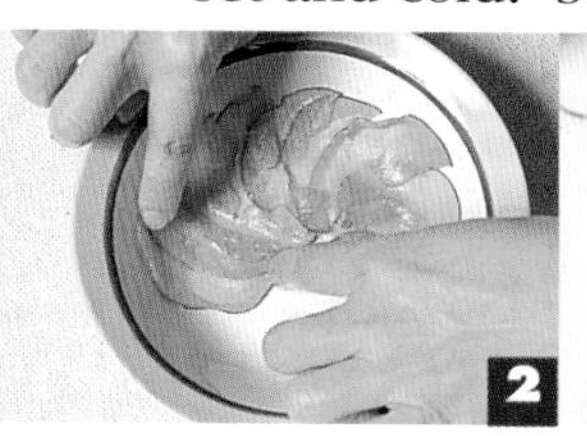
2

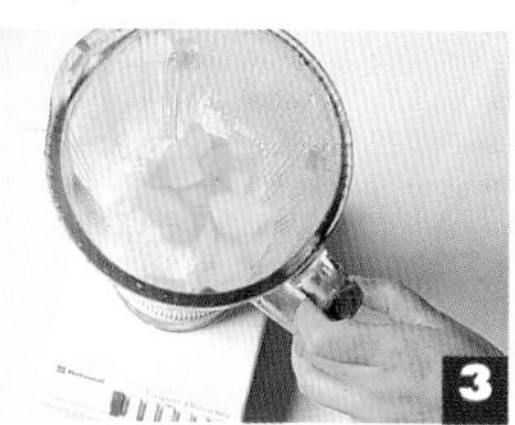
3

榴槤慕斯

Durian Mousse

INGREDIENTS:

200g. durian flesh, 1C. coconut milk, 1C. milk, 8 slices gelatin, 120g. sugar, 1C. whipped cream

METHODS:

1 Blend durian flesh and coconut milk in blender until well mashed.

2 Soak gelatin slice by slice for about 10 minutes until soft.

3 Add sugar to milk, cook over low heat until sugar dissolves. Add gelatin and stir until dissolved completely. Then add mashed durian (fig.2).

4 In a double-boiler with ice water (fig.3), stir until thickened, fold in whipped cream, and spread evenly in mold. Let cool, refrigerate until cold and set. Serve.

材料

榴槤肉200公克、椰漿1杯、鮮奶1杯、吉利丁片8片、糖120公克、打發鮮奶油約1杯

做法

1 榴槤肉加入椰漿（圖1），放入果汁機內打成泥備用。

2 吉利丁片1片片放入冰水內浸泡至軟，約10分鐘。

3 鮮奶加入糖，以小火煮至糖化，放入吉利丁片攪拌至完全融化，再倒入打好的榴槤泥中（圖2）。

4 以隔冰水法攪拌（圖3）至成稠狀後，拌入打好的鮮奶油裝入模內，待涼，放入冰箱冷藏至凝固且完全涼透。

NOTES: 你也可以換成→吉利丁粉
換成吉利丁粉約需使用1 1/2大匙。
You can substitute 1 1/2T. gelatin powder for 8 slices of gelatin.

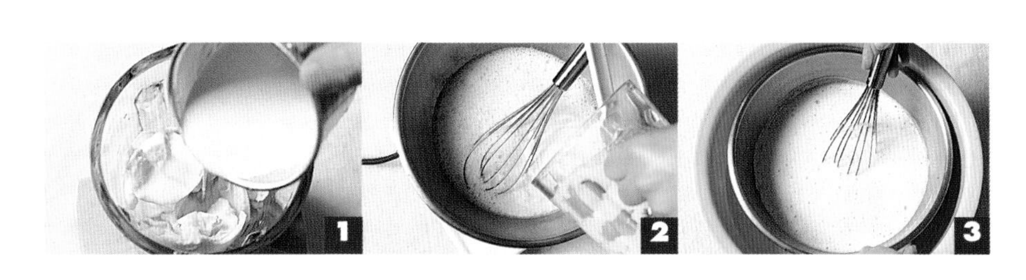

巧克力慕斯

Chocolate Mousse

材料

可可粉$1^1/_2$大匙、雞蛋2個、鮮奶1/2杯、
吉利丁粉$1^1/_2$大匙、糖60公克、熱水1/4杯、
戚風蛋糕1公分厚1片（請見P.86）

做法

1 可可粉加入鮮奶攪拌均匀，以小火煮至有香味。

2 吉利丁粉與熱水調勻備用，蛋白與蛋黃分開，蛋黃放入可可糊內攪拌均勻，再拌入吉利丁液（圖1）。

3 蛋白打至起泡後，加入糖打至糖完全溶化，且呈濕性發泡，拌入巧克力糊內（圖2）攪拌均勻。

4 取長方形模子，放入巧克力糊，再壓上1片戚風蛋糕（圖3），放入冰箱冰至完全凝固後扣出即可。

INGREDIENTS:

$1^1/_2$T. cocoa powder, 2 eggs,
1/2C. milk, $1^1/_2$T. gelatin powder,
60g. sugar, 1/4C. hot water,
1 piece 1-cm thick chiffon cake(see p.86)

METHODS:

1 Add milk to cocoa powder, stir well, cook over low heat until flavor is released.

2 Mix gelatin powder well with hot water, and set aside. Separate egg whites from yolks, and add yolks to cocoa batter. Stir until well-mixed, then add gelatin water (fig.1).

3 Beat egg whites until fluffy, add sugar, beat until sugar dissolves and bubbles appear. Add to batter (fig.2), and stir until well blended.

4 Spread batter evenly in a rectangle mold, top with chiffon cake (fig.3), refrigerate until chill and set. Flip onto plate and serve.

NOTES: 你也可以換成→吉利丁片
換成吉利丁片約需使用8片。
You can substitute 8 slices of gelatin for $1^1/_2$T. gelatin powder.

1

2

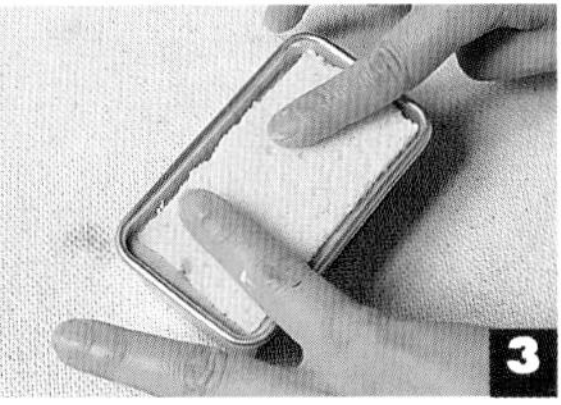
3

櫻桃乳酪慕斯

Cherry Cheese Mouse

材料

奶油乳酪150公克、櫻桃罐頭1瓶（400公克）、蛋黃2個、糖60公克、吉利丁粉1¹/₂大匙、打發鮮奶油1杯、8吋戚風蛋糕1公分厚1片（請見P.86）

做法

1. 奶油乳酪加入糖，以打蛋器打至鬆軟，再將蛋黃分兩次加入（圖1），約打至糖融化。
2. 櫻桃加入吉利丁粉（圖2），攪拌均勻後，以小火煮至吉利丁融化。
3. 櫻桃以隔冰水法攪拌至涼，但不可太濃稠；加入打好的乳酪內（圖3），攪拌均勻，再加入鮮奶油拌勻。
4. 將戚風蛋糕放在蛋糕模底部，再倒入拌好的慕斯材料，放入冰箱冷藏庫至完全凝固且涼透即可。

INGREDIENTS:

150g. cream cheese,
1 can cherries (400g), 2 egg yolks,
60g. sugar, 1-cm thick chiffon cake,
1¹/₂T. gelatin powder,
1C. whipped cream,
1 piece 8-inch large 1cm thick chiffon cake (see p.86)

METHODS:

1. Add sugar to cream cheese, beat with an egg beater until fluffy. Add yolks one at a time (fig.1), then beat until sugar dissolves.
2. Mix gelatin powder with cherries well (fig.2), cook over low heat until gelatin dissolves.
3. In double-boiler with ice water at bottom and stir until the temperature drops, add to step 1 (fig.3), stir well, add whipped cream cheese batter and mix well.
4. Line cake pan with chiffon cake, fold in mousse batter, refrigerate until set and cold. Serve.

NOTES: 你也可以換成→吉利丁片
換成吉利丁片約需使用8片。
You can substitute 8 slices of gelatin for 1¹/₂T.gelatin powder.

1

2

3

優酪乳慕斯

Yogurt Mousse

材料

優酪乳150公克、鮮奶100公克、
吉利丁片8片、蛋黃3個、糖50公克、
打發鮮奶油1 1/2杯、
戚風蛋糕1公分厚1片（請見P.86）

做法

1 吉利丁片1片片放入冰水內浸泡至軟約10分鐘。取心形慕斯模，將蛋糕壓成心形小塊。

2 優酪乳與鮮奶、糖混和放入不鏽鋼盆內，以小火煮至糖化，再放入捏乾水分的吉利丁片（圖1），至完全融化，待稍涼，加入蛋黃拌勻。

3 準備1個較大的盆子，放入冰水，再放入不鏽鋼盆，以隔冰水的方式（圖2）攪拌至溫度降低，呈稠狀後拌入1杯鮮奶油。

4 將拌好的奶糊盛入蛋糕上（圖3），放入冰箱內冰至凝固且涼透，取出模子，表面以鮮奶油裝飾。

NOTES: 你也可以換成→吉利丁粉
換成吉利丁粉約需使用1 1/2大匙。
You can substitute 1 1/2T. gelatin powder for 8 slices of gelatin.

INGREDIENTS:

150g. yogurt, 100g. milk,
8 slices gelatin, 3 yolks, 50g. sugar,
1 1/2C. whipped cream,
1 piece 1-cm thick chiffon cake (see p.86)

METHODS:

1 Soak gelatin in ice water, slice by slice until soft, for about 10 minutes. In a heart-shaped mousse pan, and press cake into small pieces.

2 Combine yogurt, milk and sugar in a stainless steel mixing bowl well. Cook over low heat until sugar dissolves, add draining gelatin (fig.1), stir until dissolved completely. Let cool, add yolks, stir until well-blended.

3 In a double-boiler with ice water at bottom, stir yogurt batter (fig.2) until temperature drops and it thickens, add 1C. whipped cream.

4 Fold batter onto cake (fig.3), refrigerate until set and cold, unmold, and garnish with whipped cream. Serve.

巧克力慕斯球

Chocolate Mousse Ball

材料

巧克力100公克、蛋黃1個、糖20公克、玉米粉20公克、鮮奶2/3杯、打發鮮奶油1/2杯、吉利丁粉1茶匙、熱水1大匙、核桃丁數粒

做法

1 巧克力隔水融化，倒入圓形模內（圖1），不停轉動至四周結一層薄薄巧克力，再將剩餘巧克力液倒出，待完全凝固後，將空心巧克力球取出。

2 蛋黃與糖、玉米粉放入不鏽鋼盆內混合拌勻（圖2），慢慢加入鮮奶（圖3）拌勻；吉利丁粉與熱水調勻備用。

3 鍋內燒2杯熱水，將盆子放上，以隔熱水方式煮成稠狀，加入吉利丁液拌勻，待涼後加入鮮奶油拌勻。

4 將煮好的餡料填入巧克力球內（圖4），上放1粒核桃丁，放入冰箱冰至凝固且涼透即可。

NOTES: 你也可以換成→吉利丁片
換成吉利丁片約需使用2片。
You can substitute 2 slices of gelatin for 1t. gelatin powder.

INGREDIENTS:

100g. chocolate, 1 egg yolk, 20g. sugar, 20g. cornstarch, 2/3C. milk, 1/2C. whipped cream, 1t. gelatin powder, 1T. hot water, walnut dice as needed

METHODS:

1 Melt chocolate in a double-boiler, pour into round molds (fig.1), coat mold with a thin layer of chocolate by turning mold until completely coated, pour out extra chocolate liquid. Wait until set, unmold the hollow chocolate balls.

2 Combine egg yolk, sugar and cornstarch in a stainless steel mixing bowl well (fig.2), add milk slowly (fig.3), mix well. Mix gelatin powder in hot water well, and set aside.

3 Bring 2C. water in pan to boil, with mixing bowl in center, double-boil until thickened. Add gelatin powder water, mix well, and cool first before adding whipped cream.

4 Stuff batter in chocolate balls (fig.4), top with 1 walnut dice, refrigerate until set and cold. Serve.

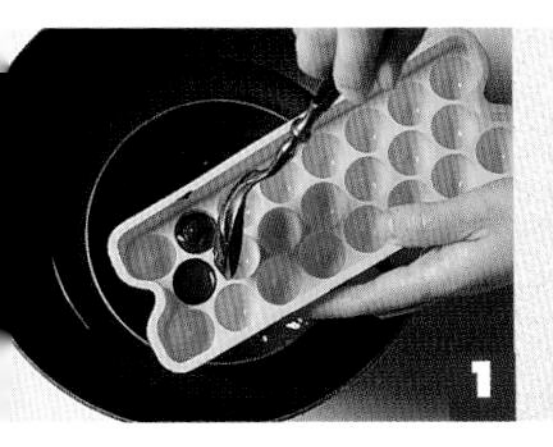
1

2

3

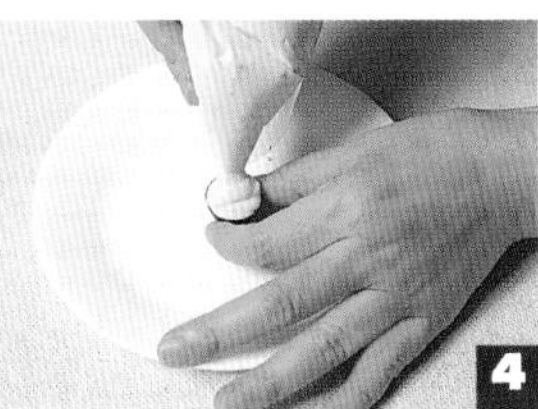
4

咖啡乳酪派

Coffee Cheese Pie

材料

奶油乳酪200公克、鮮奶1杯、糖80公克、即溶咖啡粉2大匙、熱水1/2杯、鮮奶油1杯、吉利丁片8片、8吋戚風蛋糕1公分厚1片（請見P.86）

做法

1 奶油乳酪壓碎成小塊，加入鮮奶、糖，放在爐火上（圖1），以小火煮至完全融化。

2 咖啡粉調入熱水後，倒入乳酪糊中（圖2），攪拌均勻。

3 吉利丁片1片片放入冰水中，捏乾水分後放入咖啡溶液內（圖3），至完全融化。

4 準備1個較大盆子，內放冰水，以隔水法將咖啡溶液攪拌至濃稠後，拌入鮮奶油。

5 戚風蛋糕放入蛋糕模中，倒入咖啡乳酪糊（圖4），放入冰箱冷藏庫內，冰至凝固且涼透，約需2小時。

NOTES: 你也可以換成→吉利丁粉
換成吉利丁粉約需使用$1^1/_2$大匙。
You can substitute $1^1/_2$T. gelatin powder for 8 slices of gelatin.

INGREDIENTS:

200g. cream cheese, 1C. milk, 80g. sugar, 2T. instant coffee powder, 1/2C. hot water, 1C. fresh cream, 8 slices gelatin, 1 piece 8-inch large 1-cm thick chiffon cake (see p.86)

METHODS:

1 Press cheese into small pieces, add milk and sugar, cook over low heat (fig.1) until dissolved.

2 Mix coffee powder in hot water, pour into cheese batter (fig.2), stir evenly.

3 Soak gelatin slice by slice at a time, remove and squeeze out excess water, add to coffee (fig.3), and stir until dissolved.

4 Put ice water in large mixing bowl, in a double-boiler, stir coffee mixture until thickened, fold in whipping cream.

5 Line pie pan with chiffon cake, pour in batter (fig.4), refrigerate for about 2 hours until set and cold. Serve.

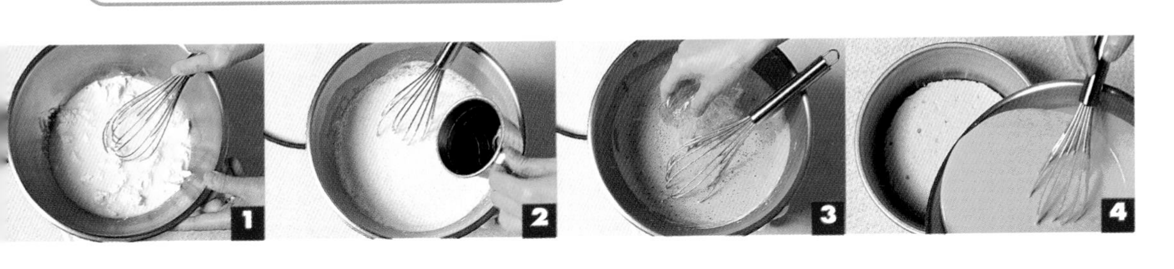

1
2
3
4

水果派

Fruit Pie

材料

蘇打餅干約10片、融化奶油60公克、蛋黃2個、鮮奶1杯、玉米粉30公克、糖80公克、吉利丁粉1大匙、熱水1/4杯、打發鮮奶油1杯、奇異果2個、鳳梨片1片、聚力T1/2大匙

做法

1. 蘇打餅干裝進塑膠袋中，再以擀麵棍擀碎（圖1），越細越好，拌入融化的奶油後填入派盤，壓緊。
2. 蛋黃與玉米粉及40公克糖拌勻（圖2），慢慢調入鮮奶，攪拌均勻。
3. 將蛋奶液放入不鏽鋼盆內，以隔水方式，放在爐上煮至呈麵糊狀（圖3）。
4. 吉利丁與熱水調勻，拌入煮好的奶糊內，攪拌至勻，待涼再拌入鮮奶油，倒入餅干派皮內（圖4），表面鋪上奇異果及鳳梨片。
5. 聚力T調勻剩餘的糖，再拌入1/2杯冷水調勻後，以小火煮至完全溶化成透明狀，待稍涼，刷在水果表面即可。

＊餅干填入派盤前，最好先在盤底鋪上一層保鮮膜，方便將做好的派由盤內取出。

NOTES: 你也可以換成→吉利丁片
以吉利丁片換成吉利丁粉，則須使用約5片。
以吉利丁片換成聚力T，則須使用約2 1/2片。
You can substitute 5 slices of gelatin for 1T. gelatin powder.
Or You can substitute 2 1/2 slices of gelatin for 1/2T. jelly T.

INGREDIENTS:

about 10 soda crackers,
60g. melt butter, 2 egg yolks, 1C. milk,
30g. cornstarch, 80g. sugar,
1T. gelatin powder, 1C. whipped cream,
2 kiwis, 1 slice pineapple, 1/2T. jelly T

METHODS:

1. To make the pie shell：put crackers in plastic bag, crush with a rolling pin (fig.1), (the finer the better), add butter, mix well and stuff in pie mold, press tightly to form pie shell.
2. Combine egg yolks, cornstarch and half of the sugar well (fig.2), pour in milk slowly, stir constantly until well mixed.
3. Remove step 2 to a stainless steel mixing bowl, cook in double-boiler until it forms batter (fig.3).
4. Mix jelly T with hot water, add batter, stir until evenly mixed, cool first, then add whipped cream, mix well, pour in shell (fig.4), garnish with kiwi and pineapple slices.
5. Mix jelly T with rest of sugar, then add 1/2C cold water, stir well, cook over low heat until dissolved completely and transparent. Let cool and brush jelly T syrup on sliced fruit to form a layer.

＊Line pie pan with saran wrap before making shell enable you to remove pie more easily later.

芋泥派

Taro Pie

材料

去皮芋頭300公克、鮮奶1杯、糖60公克、蛋白2個、打發鮮奶油$1\frac{1}{2}$杯、吉利丁粉1大匙、熱水3大匙、烤好8吋派皮1個（請見P.88）

做法

1. 芋頭切薄片，以大火蒸約25分鐘至熟，取出放入盆內，加入20公克糖，以攪拌器打成泥，或以湯匙壓成泥，並慢慢加入鮮奶（圖1）。
2. 將芋泥與1杯鮮奶油徹底攪拌均勻（圖2），吉利丁與熱水調勻，加入芋泥中。
3. 蛋白打至起泡，與剩餘的40公克糖（圖3）打至挺立，拌入芋泥中。
4. 將調好的芋泥填入涼透的派皮中，放入冰箱中至完全凝固，以剩餘的鮮奶油裝飾即可。

INGREDIENTS:

300g. peeled taro, 1C. milk, 60g. sugar, 2 egg. whites, $1\frac{1}{2}$C. whipped cream, 1T. gelatin powder, 3T. hot water, 1 baked 8-inch pie crust (See p.88)

METHODS:

1. Slice taro, cook on high for about 25 minutes until done, remove to a large mixing bowl. Add 20g. sugar, beat with electric mixer until mashed, (or mashed by pressing with a spoon), then add milk slowly (fig.1).
2. Mix taro mash with 1C. whipped cream well (fig.2). Mix gelatin with hot water, and add to taro.
3. Beat egg whites until fluffy, add rest of 40g. sugar (fig.3), beat until mixture forms stiff peaks, fold into taro.
4. Stuff taro in pie crust, refrigerate until set, garnish with remaining whipping cream. Serve.

芒果塔

Mango Tart

材料

芒果肉150公克、糖40公克、蛋黃2個、打發鮮奶油1杯、吉利丁粉1大匙、熱水1/4杯、烤好塔皮約20個（請見P.88）

【裝飾材料】

鮮奶油適量、芒果1個、櫻桃5個

做法

1 芒果肉與糖混和打成泥（圖1）。

2 吉利丁與熱水調勻，放入芒果泥內（圖2）。

3 加入蛋黃，攪拌均勻，待涼後再拌入鮮奶油（圖3）攪拌均勻，以隔水法攪至呈稠狀，填入塔皮內。

4 表面以鮮奶油、芒果、櫻桃裝飾。

INGREDIENTS:

150g. mango flesh, 40g. sugar, 2 egg yolks, 1C. whipped cream, 1T. gelatin powder, 1/4C. hot water, about 20 baked tarts (See p. 88)

【GARNISH】

fresh cream as needed, 1 mango, 5 cherries

METHODS:

1 Blend mango and sugar until well-mashed (fig.1).

2 Mix gelatin and hot water well, add to mango (fig.2).

3 Add yolks, stir well, cool, fold in whipped cream (fig.3), and mix well. Cook in double-boiler until thickened, then stuff in tarts.

4 Garnish with fresh cream, mango and cherry. Serve.

NOTES: 你也可以換成→吉利丁片

若以吉利丁片換成吉利丁粉，則須使用約5片吉利丁片。

You can substitute 5 slices of gelatin for 1T. gelatin powder.

1

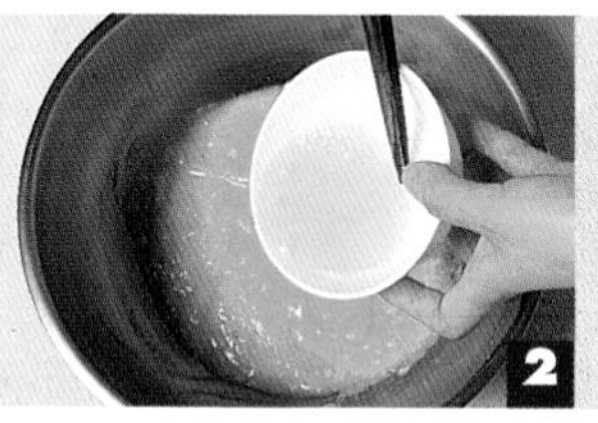

2

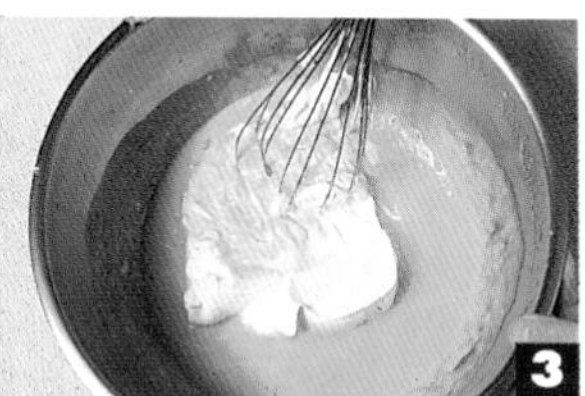

3

戚風蛋糕做法

How to Make Chiffon Cake

METHODS:

1 Combine flour and baking powder then sieve (fig.1).

2 Place egg yolks and whites in separate bowls, add 60g. sugar to yolks, stir to mix well, add evaported milk, stir gently to prevent bubbles from forming until evenly mixed (fig.2)

3 Sieve flour and baking powder, add to yolks along with vanilla essence and soy oil (fig.3), or add melted butter instead.

4 Beat whites with electric egg beater for 1 minute, add remaining sugar, beat until stiff. (fig.4)

材料

低筋麵粉110公克、泡打粉1/4茶匙、細粒砂糖120公克、雞蛋5個（連殼重300公克）、奶水1/4杯、沙拉油1/4杯、香草精1/4茶匙

做法

1 麵粉與泡打粉混合過篩。

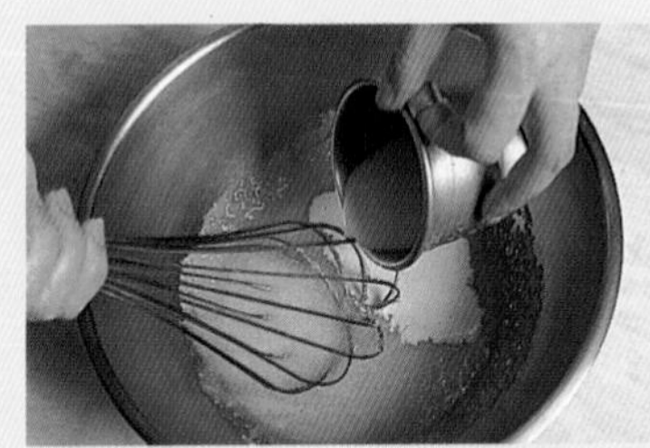

2 蛋白與蛋黃分別放在兩個盆內，取60公克糖放在蛋黃中拌勻，再加入奶水攪拌均勻，拌時不須太大力，以免產生太多氣泡 。

3 再拌入過篩的麵粉、泡打粉，拌勻後加入香草精，再加入沙拉油拌勻，可使用各種食用液體油，也可使用融化的奶油。

4 蛋白以電動打蛋器打至起泡後，約需1分鐘的時間，再加入剩餘的糖，打至以打蛋器挑起時不會垂下的程度。

INGREDIENTS:

110g. cake flour, 1/4t. baking powder,
120g. fine granulated sugar,
5 eggs (about 300g. with shells), 1/4C. evaported milk,
1/4C. soy oil, 1/4t. vanilla essence

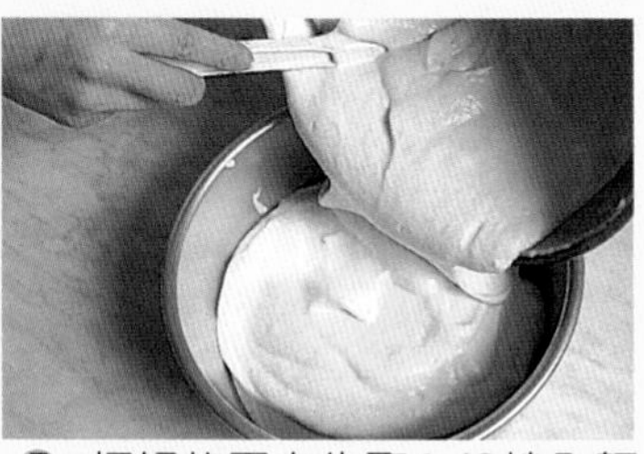

5 打好的蛋白先取1/2拌入麵糊內，大概攪拌幾下後，再拌入剩餘的蛋白，攪拌均勻。最後將拌好的麵糊倒入烤模內。

6 放入預熱好的烤箱，125°C (250°F)烤約1小時，取出時用力敲打一下，再將蛋糕扣在倒扣架上。

7 待蛋糕涼後，取下倒扣架，以抹刀由烤模邊緣劃一圈，再將蛋糕取出，並以抹刀將底盤取下。

5 Fold in half of beaten whites to yolk batter, stir gently for several times, fold in the rest, stir until evenly mixed, pour batter into baking pan (fig.5).

6 Bake in preheated oven at 125°C (250°F) for about 1 hour, remove, tap hard on table to settle contents, place upside down on rack (fig.6).

7 Cool, remove, scrape alongside pan with spatula to free cake (fig.7), unmold cake, remove bottom of pan from cake with spatula. Ready to serve.

派皮的做法

How to Make Pie Crust

METHODS:

1 Sieve flour onto kneading board, add shortening, mince shortening into tiny pieces (fig.1), coat each piece evenly with flour.

2 Dig a hole in flour, mix water and salt well, then pour into the hole (fig.2).

3 Knead into dough by working from edges (fig.3), and surround with flour.Rub the dough slightly.

4 Cover dough with cloth or saran wrap (fig.4) to keep from drying, let rise for about 20 minutes.

材料

低筋麵粉180公克、白油100公克、水1/2杯、鹽1/4茶匙。

做法

1 麵粉過篩後放在工作檯上，放上白油，以麵刀將白油切細至看不到粗粒的白油，使碎細的白油都能均匀的沾裹上麵粉。

2 將麵粉中間挖一麵牆，水與鹽調匀後，倒在中間。

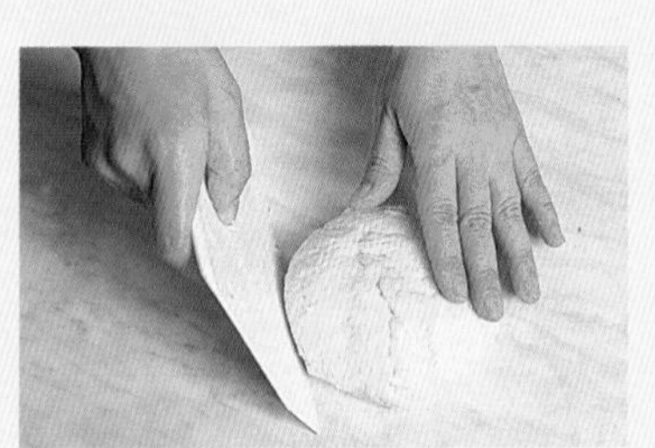

3 以手指由中心開始往外繞，至成一麵糰，外圍都要有乾粉包圍，拌時不要用力搓揉。

4 拌好後以布或保鮮膜蓋好，防止表面風乾，放置醒約20分鐘。

INGREDIENTS:

180g. cake flour, 100g. shortening, 1/2C. water, 1/4t. salt

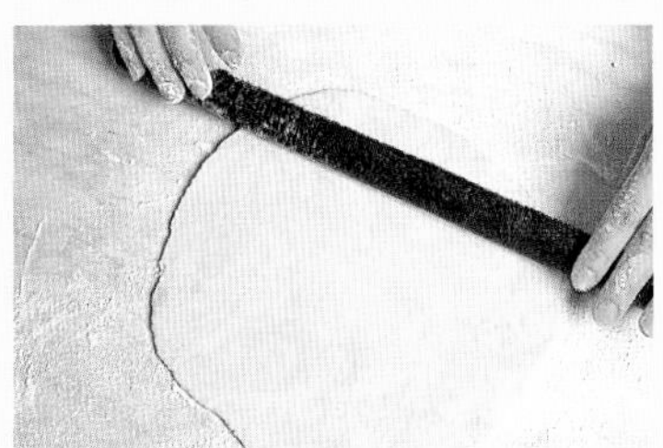

5 麵糰醒好後，在工作檯上撒上適量的高筋麵粉當手粉，將麵糰擀開至所需要的厚薄及大小。

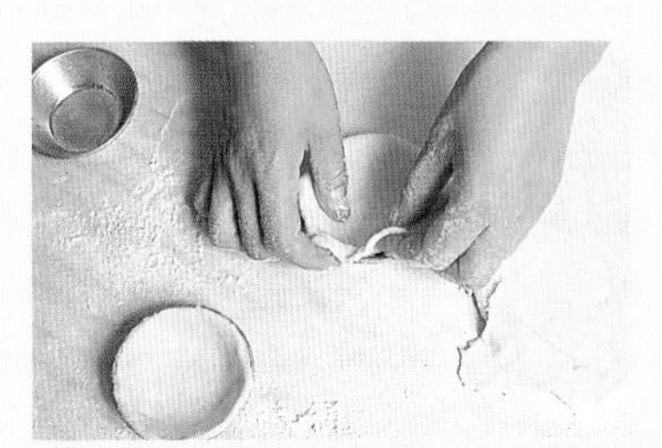

6 擀好的派皮直接填入抹油的塔模或派模內。

7 剩下來破碎的麵皮不需搓揉，只需將其合起來，堆成一堆，再醒約10分鐘，表面拍上少許手粉，即可再擀開使用。

5 Sprinkle suitable amount strong flour on kneading board, roll dough into desired size and thickness (fig.5).

6 Transfer pie crust dough into a greased tart pan or pie pan (fig.6).

7 Trim off excess and gently knead (fig.7), let rest for 10 minutes, pat with a little flour, roll out into desired size again.

CHICKEN JELLO

Vegetarian Meat Jello

TOFU WITH CORN JELLO

沁涼小菜

CHILLED SIDE DISH

LLO PEPPER SALMONJ ELLO SEAFOOD JELLO COLD SQUID ROLLS TOFU WITH CORN JELLO SPINACH WITH TOFU JELLO

玉米豆腐凍

Tofu with Corn Jello

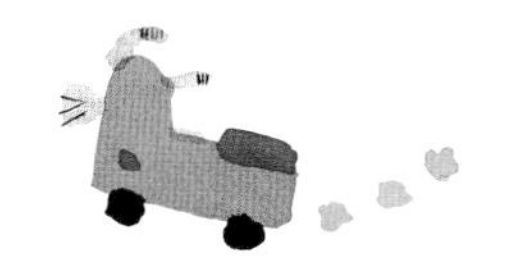

INGREDIENTS:

1C. canned corn kernels,
1 square tender tofu,
2 sheets nori wrappers, 1t. sugar,
2T. jelly T, 600c.c. water,
1/2t. salt

METHODS:

1 Drain corn kernels. Dice tofu. Cut nori sheets into 1.5cm wide strips.

2 Combine jelly T, sugar and salt well. Add water and jelly T in pan, cook over low heat until dissolved and transparent.

3 Add corn kernels and tofu (fig.2), bring to boil, remove from heat, and pour in flat pan, Let cool, refrigerate until cold and set.

4 Remove, cut into small pieces, wrap up in nori sheets (fig.3). Serve.

材料

罐頭玉米粒1杯、嫩豆腐1塊、
包壽司用紫菜2張、糖1茶匙、聚力T2大匙、
水600C.C.、鹽1/2茶匙

做法

1 將罐頭玉米粒內的水濾乾、豆腐切小丁、紫菜切約1.5公分寬。

2 聚力T與糖、鹽攪拌均勻，鍋內放入水，再加入聚力T（圖1），以小火煮至完全融化且呈透明狀。

3 加入玉米粒及豆腐（圖2），煮滾後熄火，倒入平盤中，待涼後放入冰箱冷藏庫冰至凝固且完全涼透。

4 取出切小塊，中間包上紫菜（圖3）即可。

NOTES:

此道食譜為素食，所以不可使用以動物膠製成的吉利丁片或粉來代替。
Do not substitute gelatin slices or gelatin powder for jelly T because the jello is made especially for vegetarians.

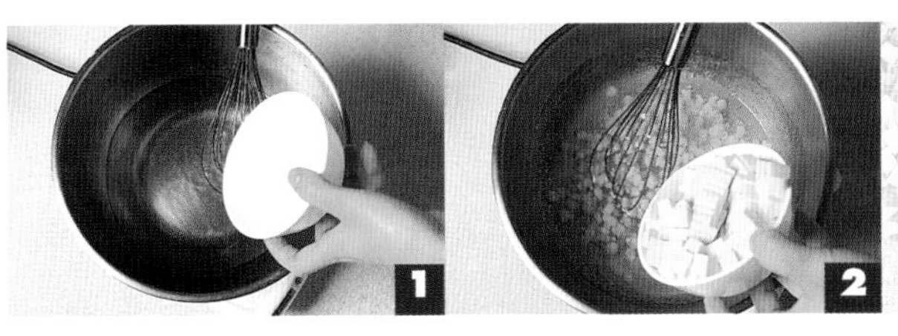

素肉凍

Vegetarian Meat Jello

材料

素肉塊30公克、小香菇約20朵、薑3片、醬油3大匙、糖2茶匙、聚力T2大匙

做法

1 素肉塊以水浸泡約1小時至軟，放入熱水內汆燙過（圖1），再以冷水沖涼後捏乾水份。

2 香菇以水泡軟，洗淨去蒂；鍋內放入2大匙油燒熱，加入香菇、薑片炒香（圖2）。

3 加入素肉塊（圖3）拌炒均勻，再加入糖、醬油及5杯水，煮至滾，改小火繼續燜煮約30分鐘，至約剩2杯湯汁。

4 聚力T加入1/2杯水調勻，倒入肉塊內（圖4），煮至完全融化，再將煮好的材料倒入模內，待涼後放入冰箱冷藏，至完全冰涼（約2小時），扣出後即可食用。

INGREDIENTS:

30 g. vegetarian meat chunks,
about 20 small dried black mushrooms (shiitake),
3 ginger slices, 2t. sugar , 2T. jelly T

Methods:

1 Soak vegetarian meat in water for about 1 hour until soft, blanch in boiling water (fig.1), remove immediately, rinse under cold water, then squeeze out excess water.

2 Soak mushrooms in water until soft, rinse and remove stems; heat 2T. cooking oil in wok, stir-fry mushrooms and ginger until fragrant (fig.2).

3 Add vegetarian meat chunks to mushrooms (fig.3), stir constantly until evenly mixed, add sugar, soy sauce and 5 cups of water, cook until boiled, reduce heat to low, continue to simmer for about 30 minutes until there are 2C. of liquid left.

4 Combine jelly T well with 1/2C. of water, add to vegetarian meat (fig.4), cook until completely dissolved, remove cooked ingredients to mold. Let cool, remove to refrigerator and chill for about 2 hours until thoroughly cold and set, remove, flip over onto plate. Serve.

翡翠豆腐

Spinach Jello

材料

菠菜葉（只取綠葉部分）150公克、布丁粉1包、糖50公克、鹽1/2茶匙、水6杯

【1塊翡翠豆腐的調味料】

柴魚片1包、辣椒末1大匙、蔥花1大匙、醬油2大匙、香油1/4茶匙

做法

1. 菠菜葉摘小段，放入果汁機內，加入2杯水，加蓋（圖1）攪打成泥。
2. 將菠菜汁過濾，去除殘渣（圖2）。
3. 鍋內放入剩餘的4杯水，加入菠菜汁拌勻；糖、鹽及布丁粉攪拌均勻後，倒入菠菜汁內（圖3），攪拌均勻，以小火煮至滾，且布丁粉完全融化。
4. 稍涼後盛入模型內，至涼後放入冰箱冷藏，完全涼透後扣出。
5. 上撒柴魚片、辣椒末、蔥花，食用時淋上醬油、香油。

NOTES: 你也可以換成→聚力T
換成聚力T約需使用5大匙。
You can substitute 5T. jelly T for pudding powder.

INGREDIENTS:

150g. spinach leaves (green leaves only), 1 pack pudding powder, 50g. sugar, 1/2t. salt, 6C. water

【SEASONING (for 1 spinach jello)】

1 pack bonito flakes,
1T. minced chilli pepper,
1T. chopped scallion, 2T. soy sauce,
1/4t. sesame seed oil

METHODS:

1. Tear spinach leaves into small pieces, blend in blender with 2C. water until well-mashed (fig.1).
2. Sieve spinach liquid, and discard dregs (fig.2).
3. With 4C. water in pan, add spinach liquid, and mix well. Mix sugar, salt and pudding powder well, add to spinach water (fig.3). Cook over low heat until boiled and pudding powder dissolves completely.
4. Cool, remove to mold. Refrigerate until cold, remove, and flip onto plate.
5. Garnish with bonito flakes, chilli pepper and scallions. Drip soy sauce and sesame seed oil on top. Serve.

虎皮凍

Tiger Skin Jello

材料

豬皮500公克、水5杯、蔥2根、薑3片、酒2大匙、醬油1大匙、鹽1/4茶匙

【蘸醃醬材料】

醬油2大匙、糖1大匙、醋1大匙、蒜泥1大匙、辣椒末1茶匙、香油1/4大匙

做法

1 豬皮洗淨，處理乾淨（請見P.7）；蔥切段，薑切片。

2 鍋內加入水，放入豬皮、蔥段、薑片、酒，煮滾後改小火燜煮約50分鐘，取出豬皮，切成約0.5公分寬的粗絲（圖1），再放入豬皮、拿掉蔥薑，繼續煮約10分鐘，至豬皮軟，可用筷子挾斷的程度。

3 加入醬油（圖2）及鹽，倒入長方形模型內，待涼，放入冰箱冷藏至完全凝固。

4 調勻蘸醬材料，食用時將豬皮凍切薄片，配上蘸料即可。

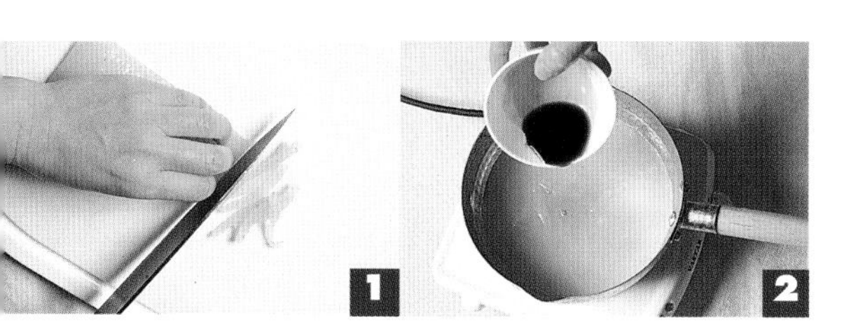

INGREDIENTS:

500g. pork skin, 5C. water, 2 scallions, 3 ginger slices, 2T. cooking wine, 1T. soy sauce, 1/4t. salt

【SEASONING】

2T. soy sauce, 1T. sugar, 1T. vinegar, 1T. mashed garlic, 1t. minced chilipepper, 1/4T. sesame oil

METHODS:

1 Rinse pork skin, clean thoroughly (See p.9). Cut scallions into sections and slice ginger.

2 Pour water in pan, add pork skin, scallion, ginger and wine, bring to boil, then reduce heat to low and simmer for 50 minutes. Remove pork skin, shred into 0.5cm (fig.1) strips, return to pan, discard scallions and ginger, continue to cook for 10 minutes until soft and easily broken by chopsticks.

3 Add soy sauce (fig.2) and salt, pour in rectangular pan. Let cool and refrigerate until set.

4 Combine ingredients of seasoning well, slice jello before serving and drip with the seasoning.

雞凍

Chicken Jello

材料

雞腿2根（約500公克）、蔥4根、薑6片、豬皮約200公克、醬油2大匙、糖2茶匙、香油1/4茶匙、鹽1/4茶匙

做法

1. 雞腿去骨後切小塊，蔥切斜段。
2. 豬皮以熱水汆燙後，將內部之油徹底刮除乾淨，加入2根蔥及3片薑，再加入約4杯水，煮滾後改小火慢慢熬煮約50分鐘，將豬皮取出切絲。
3. 鍋中熱油2大匙，放入剩下的蔥薑炒香，再加入雞腿炒至約八分熟（圖1），加入糖及醬油拌炒均勻後，最後加入煮好的豬皮湯汁（圖2），繼續煮約20分鐘，將蔥薑拿掉不要，再加入鹽及香油調味。
4. 取一碗湯，將煮好的雞肉放入（圖3），待涼，放入冰箱冷藏至凝固且涼透即可食用。

NOTES: 你也可以將豬皮換成→吉利丁粉or吉利丁片
吉利丁粉的份量以雞肉內的湯汁多寡來決定，若雞肉煮好後有2杯湯汁，則須2大匙吉利丁粉或10片吉利丁片
You can substitute gelatin powder or gelatin slices for pork skin. According to the amount of chicken soup to decide how much gelatin should be added. Add 2T. gelatin or 10 slices of gelatin if it remains 2 cups of soup after the chicken is done.

INGREDIENTS:

2 chicken legs (about 500g.), 4 scallions, 6 ginger slices, about 200g. pork skin, 2T. soy sauce, 2t. sugar, 1/4t. sesame seed oil, 1/4t. salt

METHODS:

1. Remove bones from chicken, dice; cut scallions into sections diagonally.
2. Blanch pork skin in hot water, remove, clean inner side thoroughly, remove extra fat, cook in 4C. water with 2 scallions and 3 ginger slices, bring to boil first, then reduce heat to low and cook for about 50 minutes, remove pork skin and shred.
3. Heat 2T. cooking oil in wok, stir-fry rest of scallions and ginger slices until fragrant, add chicken, stir-fry until medium done (fig.1), add sugar and soy sauce to taste, stir until evenly mixed, add liquid from cooking pork skin (fig.2), continue to cook for about 20 minutes. Discard the scallion and ginger, add salt and pepper to taste.
4. Prepare a bowl of soup, add chicken (fig.3), let cool. Remove to refrigerator and chill until set and cold. Serve.

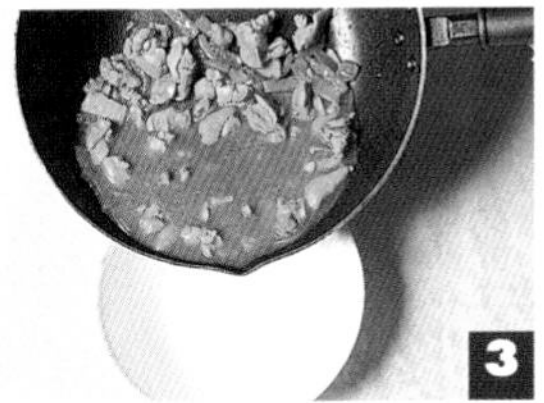

牛肉凍

Beef Jello

INGREDIENTS:

300g. beef fillet, 200g. pork skin,
2 scallions, 3 ginger slices,
2T. soy sauce, 2t. sugar,
1/4t. sesame seed oil

METHODS:

1 Rinse pork (See p.9 for cleaning and cooking), cook until white and thick (fig.1).

2 Dice beef. Cut scallions diagonally into sections. Heat 2T. cooking oil in wok, stir-fry scallions and ginger until fragrant, add beef, stir evenly (fig.2).

3 Add pork skin and about 2C. liquid from cooking pork skin (fig.3), add sugar and soy sauce, continue to cook for 10 minutes.

4 Discard pork skin, scallions and ginger, pour liquid into mold. Let cool, refrigerate until set and cold. Serve.

材料

牛腓力300公克、豬皮約200公克、蔥2根、薑3片、醬油2大匙、糖2茶匙、香油1/4茶匙

做法

1 豬皮洗淨（處理及熬法請見P7），熬至湯汁呈乳白色（圖1）。

2 牛肉切丁、蔥切斜段、鍋內熱油2大匙，放入蔥薑爆香，再加入牛肉拌炒（圖2）。

3 加入煮好的豬皮（鍋中加入2杯湯汁）（圖3）及糖、醬油，繼續煮約10分鐘。

4 煮好後將豬皮、蔥、薑挑出，其餘倒入模型中，待涼，放入冰箱，冷藏至凝固且涼透即可食用。

雞肉玉米凍

Chicken with Corn Jello

INGREDIENTS:

200g. chicken breast fillet, 1C. corn kernels, 1/4 onion, 2T. gelatin powder, 1/2t. salt, 1/4t. pepper

METHODS:

1. Add 1/2C water to gelatin powder(fig.1), mix well, and heat until dissolved.
2. Dice chicken and onion.
3. Heat 2T. cooking oil in wok, stir-fry onion until soft and fragr ant, add chicken, and fry until nearly done. Add corn kernels (fig.2) and 2C. water, bring to boil, add salt and pepper to taste.
4. Add gelatin water (fig.3), stir until well-mixed, pour in bowl. Let cool and refrigerate until set and cold. Serve.

材料

雞胸肉200公克、玉米粒1杯、洋蔥1/4個、吉利丁粉2大匙、鹽1/2茶匙、胡椒粉1/4茶匙

做法

1. 吉利丁粉加入1/2杯水（圖1），拌勻後加熱至融化。
2. 雞胸肉、洋蔥切小丁。
3. 鍋內熱油2大匙，放入洋蔥炒香至軟，加入雞肉續炒至肉變色，再加入玉米粒（圖2）及2杯水煮至滾後，加鹽及胡椒調味。
4. 加入吉利丁液（圖3）攪拌均勻後，倒入湯碗內，待涼後放入冰箱冷藏至凝固且涼透即可。

NOTES: 你也可以換成→吉利丁片or聚力T

換成吉利丁片約需使用10片。

換成聚力T約需使用1 1/2大匙。

You can substitute 10 slices of gelatin or 1 1/2T. jelly T for 2T. gelatin powder.

菠菜火腿凍

Spinach with Ham Jello

材料

火腿約300公克、波菜葉150公克、冷凍什錦蔬菜約1/2杯、高湯1杯、水1杯、洋菜8公克、鹽1/3茶匙、胡椒粉少許

做法

1 洋菜放入水內浸泡至軟，約20分鐘；火腿切丁。

2 高湯與水混合，放入捏乾水分的洋菜（圖1），煮至完全融化，加入什錦蔬菜、火腿丁，再加入鹽及胡椒粉（圖2）。

3 波菜放入熱水內汆燙至軟，取出以冷開水沖涼，捏乾水分，1片片攤平，整齊的排列在模子內（圖3）。

4 將煮好的火腿倒入菠菜模內，表面以波菜葉蓋好，待涼後放入冰箱冷藏至凝固及涼透，食用時切片即可。

INGREDIENTS:

about 300g. ham, 150g. spinach leaves, 1/2C. frozen mixed vegetables, 1C. soup broth, 1C. water, 8g. agar- agar, 1/3t. salt, pepper as needed

METHODS:

1 Soak agar- agar in water for 20 minutes until soft. Dice ham.

2 Combine broth and water, add draining agar-agar (fig.1), cook until dissolved, add mixed vegetables and ham(fig.2), add salt and pepper.

3 Blanch spinach in hot water until soft, remove, rinse under running cold water, drain spread out leaf by leaf, and arrange on bottom of pan (fig.3).

4 Pour ham mixture into pan, cover with spinach leaves. Let cool, and refrigerate until cold and set. Slice and serve.

NOTES: 你也可以換成→吉利丁片、吉利丁粉or聚力T

換成吉利丁片約需使用10片。

換成吉利丁粉約需使用2大匙。

換成聚力T約需使用1 1/2大匙。

You can substitute 10 slices gelatin, 2T.gelatin powder, or 1 1/2T. jelly T for 8g. agar-agar.

甜椒鮭魚凍

Pepper Salmon Jello

材料

鮭魚150公克、甜椒2個、玉米粒2大匙、青豆仁1大匙、胡椒粉少許、鹽1/2茶匙、糖1茶匙、吉利丁粉1大匙、熱水1杯

做法

1 鮭魚蒸熟，或以微波爐微波約3分鐘取出切小丁（圖1）。

2 甜椒由蒂部往下約0.5公分之位置橫切開，再將內部之瓜瓤挖除；同時將切下之蓋子切除蒂後切小丁。

3 將甜椒放入熱水內，以小火煮約3分鐘（圖2），取出以冷水沖涼。

4 吉利丁與熱水調勻備用，鍋內熱油1大匙，放入切丁的甜椒、玉米粒及青豆仁炒至熟，加入鮭魚，再加入鹽、胡椒粉及糖調味，再倒入調好的吉利丁液攪拌均勻。

5 將煮好的材料填入甜椒內（圖3），待涼後放入冰箱冷藏至凝固，涼透後切片即可食用。

NOTES: 你也可以換成→吉利丁片
換成吉利丁片約需使用5片
You can substitute 5 slices of gelatin for 1T.gelatin powder.

INGREDIENTS:

150g. salmon, 2 peppers, 2T. corn kernels, 1T. peas, pepper as needed, 1/2t. salt, 1t. sugar, 1T. gelatin powder, 1C. hot water

METHODS:

1 Steam salmon until done, or cook in microwave for 3 minutes, remove and dice (fig.1).

2 Cut pepper horizontally about 0.5cm deep from stem, remove inner flesh and seeds, discard stem and dice the top.

3 Cook pepper in hot water over low heat for about 3 minutes (fig.2), remove and rinse under cold water.

4 Mix gelatin with hot water well; heat 1T. cooking oil in wok, stir-fry diced pepper, corn kernels and peas until done, add salmon, then add salt, pepper and sugar to taste, pour in gelatin water, stir until evenly mixed.

5 Stuff step 4 in peppers (fig.3), let cool, remove to refrigerator and chill until set and cold, remove, slice and serve.

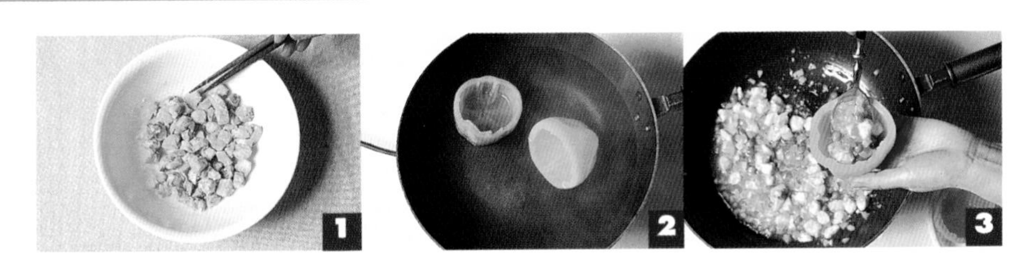

石榴海鮮凍

Seafood Jello

材料

蝦仁200公克、透抽150公克、高湯1杯、
玉米粒2大匙、青豆仁1大匙、鹽1/2茶匙、
胡椒粉少許、吉利丁片10片、
10公分見方玻璃紙12張

做法

1 蝦仁剔除腸泥洗淨，透抽切小丁，將蝦仁、透抽放入熱水內汆燙至熟，取出以冷開水沖涼。

2 吉利丁片以冰水浸泡約10分鐘備用。

3 鍋內放入高湯及1杯水，煮滾後加入吉利丁片，再拌入蝦仁、透抽、玉米粒、青豆仁（圖1），以鹽及胡椒粉調味。

4 取一小湯碗，放上玻璃紙，再放入煮好的材料（圖2）。

5 將玻璃紙往中心拉高，再以繩子綁緊（圖3），待材料涼後，放入冰箱冷藏至凝固，且完全涼透即可食用。

＊使用小湯碗固定，較好包裹，且材料不易流出。

NOTES: 你也可以換成→吉利丁粉
換成吉利丁粉約需使用2大匙，吉利丁的口感較Q，所以最好不要以聚力T代替。
You can substitute 2T. gelatin powder for 10 slices of gelatin. Do not substitute jelly T for gelatin because use gelatin to make the jelly would taste chewier.

INGREDIENTS:

200g. shelled shrimp, 150g. squid,
1C. broth, 2T. corn kernels,
1T. peas, 1/2t. salt, pepper as needed,
10 slices gelatin,
12 sheets 10cmx10cm parchment paper

METHODS:

1 Remove the intestines of shrimps and rinse thoroughly, dice squid. Blanch shrimp and squid in hot water until done, remove and rinse under cold water.

2 Soak gelatin in ice water for 10 minutes.

3 Add broth and 1C. water in pan, bring to boil, add gelatin first, followed shrimp, squids, kernels and peas (fig.1), then add salt and pepper to taste.

4 Line a small bowl with parchment paper, fill with cooked seafood (fig.2).

5 Bring corners of parchment paper to center to wrap up the ingredients, tie tightly with strings(fig.3). Let cool, remove to refrigerator until set and cold completely, remove and serve.

＊Using a small soup bowls to hold the cooked ingredients makes it easier to tie up and keeps liquid from draining away.

1

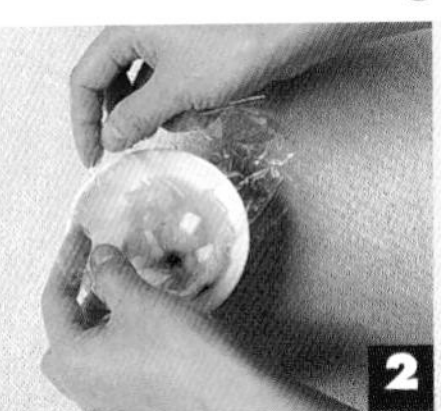
2

3

冰涼魷魚卷

Cold Squid Rolls

材料

透抽2條、豌豆仁2大匙、玉米粒2大匙、
薑片4片、吉利丁片5片、鹽1/2茶匙、水1杯

做法

1 透抽剝除外皮，並以筷子將腹內的腸泥抽出，洗淨。

2 鍋內燒熱水，放入透抽及薑片煮至熟（圖2），撈起以冷開水沖涼。

3 將透抽之頭鬚切小丁（圖2）。

4 鍋內放入1杯水，再加入透抽丁、豌豆仁、玉米粒，以小火煮至滾，加鹽調味。

5 吉利丁以冰水浸泡至軟，放入煮好的材料內，攪拌至完全融化，填入透抽腹內（圖3），冰至凝固。食用時切片即可。

INGREDIENTS:

2 squids, 2T. peas, 2T. corn kernels, 4 ginger slices, 5 slices gelatin, 1/2t. salt, 1C. water

METHODS:

1 Remove skin and internal organs from squid, rinse well.

2 Bring water in pot to boil, cook squid until done with ginger slices added (fig.1), remove and rinse under cold water.

3 Dice squid tentacles (fig.2).

4 In pan with 1C. water, add diced squid tentacles, peas and corn kernels, cook over low heat until boiled, add salt to taste.

5 Soak gelatin in ice water until soft, add to step 4. Stir until dissolved completely, stuff in squid (fig.3), remove to refrigerator, chill until set, remove and slice. Serve.

NOTES: ＊清洗透抽時不要讓底部有破洞，否則吉利丁液就會流掉。

你也可以換成→吉利丁粉

換成吉利丁粉約需使用1大匙

＊While washing the squids, remove the internal organs but keep the whole structures remained to prevent the gelatin liquid from leaking.

You can substitute 1T. gelatin powder for 5 slices of gelatin.

全省材料行

各大百貨公司的超級市場或家電用品部門
也多少有烘焙原料及工具、機具販賣喔！
出門採買前最好打電話確定一下烘焙材料行的營業時間，
才不會白跑一趟哦！

《基隆市區》

嘉美行
基隆市豐稔街130號B1
(02)2462-1963
◎烘焙原料、工具

富盛烘焙材料行
基隆市南榮路50號
(02)2425-9255
◎烘焙原料、工具、教室

美豐商店
基隆市孝一路36號
(02)2422-3200
◎烘焙原料、工具

全愛烘焙食品行
基隆市信二路158號
(02)2428-9846
◎烘焙原料、工具

証大食品原料行
基隆市七堵區明德一路247號
(02)2456-6318
器具、機具、原料

《台北市區》

京原企業有限公司
台北市承德路七段401巷971號
(02)2893-2792
◎奶油、起司原料

惠康國際食品有限公司
台北市天母北路58號
(02)2872-1708
◎烘焙原料、工具

益和商店
台北市中山北路七段39號
(02)2871-4828
◎進口食材

僑大生活百貨
台北市德行西路45號
(02)2831-5466
◎器具

大億食品材料行
台北市大南路434號
(02)2883-8158
◎烘焙原料、工具、教室

飛訊烘焙材料總匯
台北市承德路四段277巷83號
(02)2883-0000
◎烘焙原料、工具、教室

皇品食品原料行
台北市內湖路二段13號
(02)2658-5707
◎烘焙原料、工具

精緻生活館COOK&BOOK
台北市成功路四段348號
(02)8792-2989
◎食譜、烘焙原料、教室

正大行
台北市康定路3號
(02)2311-0991
◎烘焙原料、工具

洪春梅西點器具店
台北市民生西路389號
(02)2553-3859
◎烘焙原料、工具

燈燦食品有限公司
台北市民樂街125號
(02)2557-8104
◎烘焙原料、工具

白鐵號
台北市民生東路二段116號
(02)2551-3731
◎烘焙原料、工具、教室

得榮行
台北市甘州街50號1樓
(02)2555-7162
◎烘焙原料、工具

HANDS台隆手創館
台北市中華路一段88號
(02)2331-9393

◎烘焙原料、工具

福利麵包
台北市中山北路三段23-5號
(02)2594-6923
◎烘焙原料

台北仁愛路四段26號
(02)2702-1175
◎烘焙原料

向日葵烘焙材料
台北市敦化南路一段160巷16號
(02)8771-5775
◎烘焙原料、工具

禾廣有限公司
台北市延吉街131巷12號1樓
(02)2741-6625
◎進口食材

快樂商行
台北市南京東路五段41-1號
(02)2745-7921
◎烘焙原料、工具、教室

點心奶奶烘焙房
台北市三民路113巷19號
(02)2767-9610
◎烘焙原料、工具、教室

義興西點原料行
台北市富錦街578號
(02)2760-8115
◎烘焙原料、工具

申崧食品有限公司
台北市延壽街402巷2弄13號
(02)2769-7251
◎西餐、西點原料

特力和樂股份有限公司
台北市新湖三路23號
(特力廣場1樓)
(02)8791-5566
◎烘焙工具

得宏器具原料專賣店
台北市研究院路一段96號
(02)2783-4843
◎烘焙原料、工具、教室

源記食品有限公司
台北市崇德街146巷4號1樓
(02)2736-6376
◎烘焙原料

岱里食品事業有限公司
台北市虎林街164巷5樓1樓
(02)2725-5820
◎烘焙原料

媽咪商店
台北市師大路117巷6號
(02)2369-9868
◎烘焙原料、工具、教室

晶萊原料材行
台北市和平東路三段212巷3號
(02)2733-8086
◎烘焙原料、工具

《台北縣》

上筌食品原料行
台北縣板橋市長江路三段112號
(02)2254-6556
◎烘焙原料、工具

旺達食品有限公司
台北縣板橋市信義路165號1F
(02)2962-0114
◎烘焙原料、工具

聖寶食品商行
台北縣板橋市觀光街5號
(02)2953-8855
◎烘焙原料、工具

全成功企業公司
台北縣板橋市互助街36號
(02)2256-0252
◎機具、教室

安欣食品原料行
台北縣中和市連城路347巷6弄33號
(02)2225-0018
◎烘焙原料、工具、教室

艾佳食品原料專賣店
台北縣中和市宜安118巷14號
(02)8660-8895
◎烘焙原料、工具、教室

嘉元食品有限公司
台北縣中和市國光街189巷12弄1-1號
(02)2959-5771
◎烘焙原料

崑龍食品有限公司
台北縣三重市永福街242號
(02)2287-6020
◎烘焙原料、工具、教室

合名有限公司
台北縣三重市重新路四段214巷5弄6號
(02)2977-2578
◎烘焙原料

馥品屋食品有限公司
台北縣樹林市大安路175號1F
(02)2686-2258
◎烘焙原料、工具

嘉美烘焙食品DIY
台北縣土城鎮峰廷街41號
(02)8260-2888
◎烘焙原料、工具

《桃園縣市》

好萊塢食品原料行
桃園市民生路475號1樓
(03)333-1879
◎烘焙原料、工具、教室

做點心過生活原料行
桃園市復興路345號
(03)335-3963
◎烘焙原料、工具、教室

華源食品原料行
桃園市中正三街38號
(03)332-0178
◎烘焙原料、工具

楊老師工作室
桃園市樹仁一街150號
(03)364-4727
◎烘焙原料、工具、教室

全成永食品機械公司
桃園市裕和街37號
(03)360-3715
◎器具

艾佳食品行
桃園縣中壢市黃興街111號
(03)468-4557
◎烘焙原料、工具、教室

桃榮食品用料行
桃園縣中壢市中平路91號
(03)422-1726
◎烘焙原料、工具

陸光食品原料行
桃園縣八德市陸光1號
(03)362-9783
◎烘焙原料、工具、教室

台揚食品
桃園縣龜山鄉東萬壽路311巷2號
(03)329-1111
◎烘焙原料

《新竹、苗栗》

正大食品原料行
新竹市中華路一段193號
(035)320-786
◎烘焙原料、工具、教室

新勝食品原料行
新竹市中山路640巷102號
(035)388-628
◎烘焙原料、工具

新盛發
新竹市民權路159號
(035)323-027
◎烘焙原料、工具、教室

萬和行
新竹市東門街118號
(035)223-365
◎模具

康迪食品原料行
新竹市建華街19號
(035)208-250
◎烘焙原料、工具、教室

《台中、彰化、南投》

總信食品原料行
台中市復興路三段109-4號
(04)220-2917
◎烘焙原料、工具、教室

永誠行
台中市民生路147號
(04)224-9876
◎烘焙原料、工具

台中市精誠路317號
(04)382-7578
◎烘焙原料、工具、教室

永美製餅材料行
台中市北區健行路665號
(04)205-8587
◎烘焙原料、工具

玉記香料行
台中市向上北路170號
(04)310-7576
◎烘焙原料、工具、教室

宏偉實業有限公司
台中市中港路一段224巷2號
(04)322-0067
◎進口食材

特力和樂股份有限公司
台中市台中港路二段1-5號
(04)319-5755
◎烘焙工具

銘豐商行
台中市西屯區中清路151-25號
(04)425-9869
◎烘焙原料、工具

利生食品有限公司
台中市西屯路二段28-3號
(04)312-4339
◎烘焙原料、工具

綠之葉
台中市南屯區大墩七街57號
(04)381-4586
◎有機原料、教室

豐榮食品原料行
台中縣豐原市三豐路317號
(04)527-1831
◎烘焙原料、工具、教室

益豐食品原料
台中縣大雅鄉神林南路53號
(04)567-3112
◎烘焙原料、工具

永誠行
彰化市三福街195號
(04)724-3927
◎烘焙原料、工具、教室

王成源食品原料行
彰化市永福街14號
(04)723-9446
◎模具

金永誠食品原料行
彰化縣員林鎮光明街6號
(04)832-2811
◎烘焙原料、工具

信通
彰化縣員林鎮復興路59巷26弄12號
(04)835-4066
◎烘焙原料、工具

順興食品原料行
南投縣草屯鎮中正路586號-5
(04)933-3455
◎烘焙原料、工具

《雲嘉縣市》

彩豐食品原料行
雲林縣斗六市西平路137號
(05)534-2450
◎烘焙原料、工具

新瑞益食品原料行
雲林縣斗南鎮七賢街128號
(05)596-4025
◎烘焙原料、工具、教室

永誠
雲林縣虎尾鎮德興路96號
(05)632-7153
◎烘焙原料、工具

福美珍食品原料行
嘉義市西榮街135號
(05)222-4824
◎烘焙原料、工具、教室

新瑞益食品原料行
嘉義市新民路11號
(05)286-9545
◎烘焙原料、工具、教室

《台南縣市》

玉記香料行
台南市民權路38號
(06)222-3927
◎烘焙原料、工具

瑞益食品有限公司
台南市民族路二段303號
(06)222-8982
◎烘焙原料、工具

永昌食品原料行
台南市長榮路一段115號
(06)237-7115
◎烘焙原料、工具

上輝
台南市建平街6號
(06)297-1725
◎烘焙原料、工具、教室

上品烘焙
台南市永華一街159號
(06)299-0728
◎烘焙原料、工具、教室

特力和樂股份有限公司
台南縣永康市中華路143號
(06)313-6660
◎烘焙工具

《高雄縣市》

玉記香料行
高雄市新興區六合一路147號
(07)236-0333
◎烘焙原料、工具

正大行
高雄市新興區五福二路156號
(07)261-9852
器具、機具

全成製餅器具行
高雄市新興區中東街157號
(07)223-2516
◎器具

德興烘焙原料專賣場
高雄市三民區十全二路101號
(07)311-4311
◎烘焙原料、工具

十代有限公司
高雄市懷安街30號
(07)381-3275
◎烘焙原料

烘焙家
高雄市鼓山區慶豐街28-1號
(07)552-4425
◎烘焙原料、工具、教室

薪豐行
高雄市苓雅區福德一路75號
(07)721-3413
◎烘焙原料、工具

旺來昌食品原料行
高雄市前鎮區公正路181號
(07)713-5345
◎烘焙原料、工具、教室

順慶食品原料行
高雄縣鳳山市中山路237號
(07)746-2908
◎烘焙原料、工具

福市企業有限公司
高雄縣仁武鄉高楠村後港巷145號
(07)346-3428
◎烘焙原料、工具

啓順食品原料行
屏東市民生路79-24號
(08)723-7896
◎食品原料、模型、教室

裕軒食品原料行
屏東縣潮州鎮太平路473號
(08)788-7835
◎烘焙原料、工具、教室

《東部、離島》

欣新烘焙食品行
宜蘭市進士路85號
(039)363-114
◎烘焙原料、工具

裕順食品有限公司
宜蘭縣羅東鎮純精路60號
(039)543-429
◎烘焙原料、工具

萬客來食品原料行
花蓮市和平路440號
(038)362-628
◎烘焙原料、工具

玉記香料行
台東市漢陽路30號
(089)326-505
◎烘焙原料、工具

永誠
澎湖縣林森路63號
(069)263-381
◎烘焙原料、工具

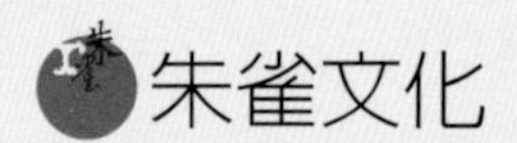

和你快樂品味休閒生活

為你精心規劃美食新生活
輕鬆做 讓你輕鬆享受烹飪樂趣

Cook50001

做西點最簡單

定價=280元 賴淑萍著

■ 蛋糕、餅干、塔、果凍、布丁、泡芙、15分鐘、簡易小點心等七大類，共50道食譜。

■ 清楚的步驟圖，就算第一次下廚也會做！詳細的基礎操作，讓初學者一看就明瞭。事前準備和工具整理，做西點絕不手忙腳亂。作者的經驗和建議，大大減少失敗機率。常用術語介紹，輕鬆進入西點世界。

Cook50002

西點麵包烘焙教室

──乙丙級烘焙食品技術士考照專書

定價=420元 乙級烘焙食品技術士陳鴻霆、吳美珠著

■ 由乙丙級技術士教導如何準備乙丙級烘焙食品技術士檢定測驗，項目：麵包及西點蛋糕。

■ 最新版烘焙食品學題庫。提供歷屆考題，每道考題均有中英文對照的品名、烘焙計算、產品製作條件、產品配方及百分比、清楚的步驟流程，以及評分要點說明、應考心得、烘焙小技巧等資訊。

Cook50005

烤箱點心百分百

定價=320元 資深烹飪老師梁淑嫈著

■ 以紮實詳細的小步驟圖帶領讀者進入西點烘焙世界，內容包括：蛋糕、麵包、派、塔、鬆餅、酥餅和餅干、小點心。教導讀者看書就會成功做點心。

■ 教你做一個師傅級的戚風蛋糕、為心愛的人裝點一個美麗的蛋糕、發麵及丹麥麵包的製作方法、千層派皮、塔皮的製作方法

Cook50008

好做又好吃的低卡點心

定價=280元 香草蛋糕鋪金一鳴著

■ 50種低熱量甜點，除了原本即屬低卡洛里的甜點外，也在傳統的甜點製作上，選用些替代的原料或不同的組合方式，讓熱愛甜點者既可盡情享受美食又不必擔心體重上升。

■ 依甜點的製作特性和材料，以春夏秋冬四季區分4大類點心，每道點心都有卡洛里數及熱量分析。

Cook50010

好做又好吃的手工麵包

──最受歡迎麵包輕鬆做

定價=320元 優仕紳麵包店陳智達著

■ 作者以從事烘焙業20年的經驗，指導讀者輕鬆做出好吃麵包的方法，包括50種最受歡迎的麵包，甜麵包、可鬆類麵包、白燒麵包、多拿滋麵包、歐式麵包、花式麵包等六大類。

■ 每單元的最開始均提供麵糰製作的過程及配方，讓讀者可直接用於同單元的麵包中，不需要做配方的換算，也不浪費麵糰。

Cook50011

做西點最快樂

定價=300元　賴淑萍著

■ 繼《做西點最簡單》，作者再推出進階級西點烘焙食譜，包括最流行的起司蛋糕、巧克力、慕斯、派、司康和瑪芬、薄餅。最熱門的提拉米蘇、義式鮮奶酪、冬日限量生巧克力、偶像日劇中的燒蘋果。

■ 日式烘焙術語解讀——輕鬆看懂日文食譜

Cook50012

心凍小品百分百

(中英對照)

定價=280元　資深烹飪老師梁淑嫈著

■ 本書運用坊間可買到的各種天然凝固劑，設計出各式各樣的甜、鹹小品。從最傳統的洋菜粉、布丁粉，到葛粉、地瓜粉，以及吉利丁、聚力T，甚至豬皮，不僅可製作甜點、果凍、冰寶，還可以做出各式各樣冰涼的菜餚。

■ 無論是夏日消暑小品或平日的開胃小點均適宜。

Cook50014

看書就會做點心

——第一次做西點就OK

定價=280元　林舜華著

■ 50種讓初學者第一次做就OK的西點。

■ 特別介紹製作西點基礎常識，如蛋白、鮮奶油打發、融化巧克力、手製擠花袋、戚風蛋糕、塔皮、派皮的做法。並列出常用的工具材料的用途、價位說明。每道西點均有作者的烹飪經驗與建議，從中學習到小技巧，減少失敗的機率。

輕鬆做001

涼涼的點心

定價=120元、特價=99元　喬媽媽著

■ 心情不好嗎？來一客清涼、溫柔的沁涼點心吧！把壞心情統統丟掉，只留下嘴裡心底酸甜的滋味。包括剉冰、蜜豆冰、雪泥等沁涼冰品和五彩繽紛的果凍及軟軟布丁。

■ 洋菜凍、吉利丁、吉利T的比較。

輕鬆做002

健康優格DIY

定價=150元　楊三連、陳小燕著

■ 帶領讀者在家自己製作衛生、高品質的優格。

■ 除各式優格冰品冷飲教學外，沾醬、濃湯、菜餚、點心，都可以加上優格，增添味覺新體驗。

■ 優格護膚小秘方，優格輕盈苗條法。關於優格的小常識及疑問解答。

國家圖書館出版預行編目

心凍小品百分百；梁淑嫈 著；施如瑛 翻譯. --初版. -- 台北市：朱雀文化，
2000〔民89〕 面； 公分. --
（Cook50；12）中英對照
ISBN 957-0309-17-2（平裝）
1.食譜 - 點心

427.16 89008085

全書圖文未經同意不得轉載和翻譯

CooK50012

心凍小品百分百

作者	梁淑嫈
攝影	張緯宇
翻譯	施如瑛
美術編輯	王佳莉
文字編輯	王美蓮
企畫統籌	李橘
發行人	莫少閒
出版者	朱雀文化事業有限公司
地址	北市建國南路二段181號8樓
電話	02-2708-4888
傳真	02-2707-4633
劃撥帳號	19234566 朱雀文化事業有限公司
e-mail	redbook@ms26.hinet.net
網址	
總經銷	展智文化事業股份有限公司
ISBN	957-0309-17-2
初版一刷	2000.07
初版二刷	2001.06
定價	280元
出版登記	北市業字第1403號

本書如有缺頁、破損、裝訂錯誤，請寄回本公司調換

Cook 50

Cook 50

Cook 50

Cook 50